# NOMOGRAPHS

## What They Are And How to Use Them

By

John Molnar

Illustrated by

Susan A. Sears

ANN ARBOR SCIENCE
PUBLISHERS INC / THE BUTTERWORTH GROUP

230 Collingwood, P.O. Box 1425, Ann Arbor, Michigan 48106

Library of Congress Card Catalog Number 81-67511
ISBN 0-250-40419-2

Manufactured in the United States of America


Butterworths, Ltd., Borough Green, Sevenoaks, Kent TN15 8PH, England

# PREFACE

This book is a collection of complementary, related nomographs covering the broad general category of problems dealing with air, water, and associated electrical and mechanical items. We have tried to include in this book those nomographs that will help you to solve rapidly and accurately the practical problems you encounter in engineering, design, maintenance and operation of systems and equipment.

These nomographs are divided into three select groups, each consisting of a general category, related to each other by the kind of problem involved. They are grouped in this manner to help you to locate quickly and easily the particular one you need. As a further convenience, we have listed each nomograph by a descriptive title in the table of contents.

## GROUP 1: AIR

This group includes those nomographs that can be used to solve problems dealing with air, properties of air and air in motion. They have one thing in common; all involve air in one manner or another.

Most of the nomographs are for air in motion, i.e., ventilation air changes, exhaust hood slot and face velocity, friction loss, heating air in motion, or transport air velocity. The others include air quantity at elevated temperature, addition and removal of moisture from air, and temperature of mixed air.

## GROUP 2: WATER

Here we have grouped those nomographs that can be used to solve problems dealing with water. This is a more diversified group than the one for air and, additionally, addresses steam.

While most nomographs are for water in motion, i.e., friction loss, velocity and quantity of water flowing in pipes, or pump motor horsepower, the other nomographs include sizing of building drains, shell and tube heat exchangers, copper and steel pipe expansion with temperature, quantity of water in steel and copper pipes, and the weight of water-filled steel and copper pipes.

## GROUP 3: ELECTRICAL, MECHANICAL AND COSTS

You will find in this group an assortment of nomographs that are associated with Groups 1 and 2 or that deal with additional problems normally found in plants and systems.

Several nomographs cover electrical power items, such as resistance, resistance/current/voltage relationships, power loss in conductors, and current/kVA and /kW relationships. The others deal with belt and pulley speeds, sheet metal duct weight and cost, power factor, concrete volume, trench excavation, electrical operating costs, simple and actual payback period, sound levels for two or more fans, and sound levels for changing fan speeds.

In these three groups we tried to include those nomographs that are useful in heating; ventilation; air conditioning; environmental control of temperature, and noise and air pollution; ducting air; piping and pumping water; and others that are related to those above that you could encounter in your daily work.

As for the nomographs themselves, we have included formulas with some where we considered it necessary or useful, and omitted formulas from others. We have included an example problem and solution in all cases so that you could work through the problem. We have excluded the graphical solution from the nomograph.

In addition to the nomographs, we have included a section on how to use the nomograph. While we readily acknowledge that this section could have been omitted for many readers of this book, we felt that some may benefit by it. For that reason we have included it, and we hope that this section will stimulate renewed interest in the use of nomographs.

We have also included a section on how to make nomographs. You may find, as we do occasionally, that there are problems for which there is no nomograph. In that case, do as we do; make your own. Even if you have never considered constructing one, with the information in this section and a little practice, you can become proficient. Also, we felt that once you have mastered the technique of constructing nomographs, you will have a better appreciation for and will be able to make greater use of the nomograph.

This book is the result of more than 30 years of engineering experience, during which time we developed many nomographs to solve our problems. Most of the nomographs in this book are ones we made and have previously appeared in major trade publications; some were specifically designed for this book, and a few select nomographs are reprinted by special permission.

The longer we worked on this book to include more and more nomographs, the more divergent the subject matter became. We decided that the quantity included covered what we originally intended. We may in the future expand this book to include more, but for the present, this is it.

We agonized greatly during the gestation of this book and hope that you will find the product worthy of the effort.

John Molnar

**John Molnar,** President of Molnar Engineering Inc., is a Professional Engineer licensed in New Jersey and Pennsylvania. He holds BS and MS degrees in engineering and is a member of the National Society of Professional Engineers.

Mr. Molnar has 30 years of experience in plant, project and consulting engineering. In addition to having published more than 100 technical articles in major trade journals, Mr. Molnar has also had an article abstracted by the Academy of Science of the U.S.S.R., Institute of Scientific Information, and has served as guest editor for *Plant Engineering* for six years. His nomograph, "Calculating the Capacity of Electric Heating Coils," was selected as The Best of Plant Engineering in 1978. His first book, *Plant Operation Manual,* was published in 1967.

Dedicated to
Elaine Naden Molnar,
for her encouragement and assistance

# CONTENTS

# INTRODUCTION

Nomographs are, simply stated, alignment charts consisting of three or more lines (scales) so arranged that with the aid of a straightedge the problem represented by the nomograph can be solved.

According to the *McGraw-Hill Dictionary of Scientific and Technical Terms*[1] a nomograph is "a chart which represents an equation containing three variables by means of three scales so that a straight line cuts the three scales in values of three variables satisfying the equation."

Nomographs are, in effect, charts with built-in calculations, used to facilitate problem solving. They are particularly useful for solving repetitive, practical problems encountered in engineering, design, maintenance and operation of facilities and systems. Once a chart is constructed for a particular equation, all that is necessary to solve a problem is a straightedge. The nomograph eliminates the probability of error in calculations because they are programmed into the charts.

Nomographs are a time-tested method of graphic computation, in fact, they have been around for a long time. Nomography, the science and technique of graphic computation, was first introduced almost a century ago by Professor Maurice d'Ocagne in France. Since then it has been developed chiefly by French and German mathematicians. The nomograph has survived the slide rule, and in these days of programmable electronic calculators, the alignment chart continues its popularity, as evidenced by its use in major industrial trade magazines.

Over the past 30 years there has been a steady increase in the construction and use of nomographs. Today nomographs are being developed to solve more complex problems than in the past. The advantage nomographs have over other methods of mathematic computation is that with one variable held constant, a continuously variable readout can be obtained by rotating the straightedge around the fixed variable. Like electronic calculators, nomographs cannot think (the user does that) but they can reel off solutions in an almost magical manner.

The type of nomograph used to solve a particular problem depends to a great extent on the complexity of the problem, and perhaps to an equal extent on the preference of the person constructing it. Given the same problem, different people will most likely use a different method and a different type of nomograph.

The form of these charts depends on the equation involved and can be a simple three parallel line chart with uniform scales or one with more than three lines, which may be a combination of parallel and nonparallel lines, straight and curved lines and uniform and nonuniform scales. Some nomographs may contain one or more pivot lines to extend their capability. A common type of nomograph is the Z (or N) chart. It is ideal for three-variable equations. A double N chart can be constructed to solve equations with four variables.

The one disadvantage of nomographs, if it can be considered as such, is that they do not give exact solutions. Fortunately, in most practical-type situations, exact numerical solutions are not required. For example, in problems dealing with water flow in pipes, it is unnecessary, and impractical to know that the pipe size required is 6.375 inches in diameter. That size is not commercially available; you will have to use either the next smaller or larger standard size. The same reasoning applies to motors, transformers, electric resistance heaters, converters, heat exchangers, drive pulleys, etc.

On the other hand, in the water flow problem, once the standard pipe size has been selected, various rates of flow in gallons per minute and velocity in feet per second can be determined by merely rotating the straightedge around the fixed pipe size selected on the nomograph.

The degree of accuracy of any nomograph depends on the size of the chart and the range of the scales. We have made the charts in this book to a size which provides reasonable accuracy, and have used scales with practical limits.

The accuracy of any engineering computation can be no greater than that of the input data, and need not be any more accurate than the solution demands. This statement applies equally to the slide rule, electronic calculator and nomograph. The nomograph is a very useful tool, even in this day of the electronic calculator, and the more sophisticated programmable calculator.

## HOW TO CONSTRUCT NOMOGRAPHS

Now through the aid of examples you will be shown how to construct the scales, locate the position of the scales and make nomographs.

Once a problem is identified and an equation written to solve it, the next step is to determine the type of nomograph that best satisfies the requirements for rapidly determining solutions. The range of the scales selected for the nomograph must satisfy the range of the problem to be solved. After the nomograph is constructed, it should then be tested solving several problems to verify the correctness of construction.

### Uniform Scales

Figure 1 shows a uniform scale; it has equally spaced division marks similar to an engineer's scale. In fact, an engineer's scale is an excellent device for constructing uniform scales, particularly the 12-in. triangular scale graduated 10-20-30-40-50-60 parts to the inch. The 50th scale was used to construct the uniform scale shown in Figure 1. It is 4 in. long, and the distances along the scale from the zero mark are equal to the numbers on the 50th scale and are merely numbers, but when used as part of a nomograph, they could represent weight, volume, velocity, etc., depending on the equation involved.

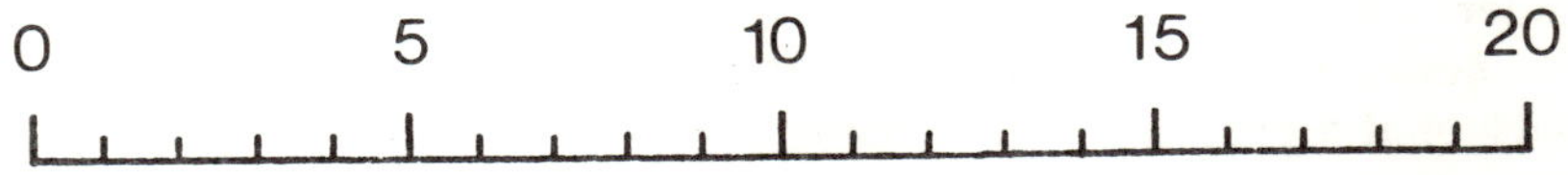


Figure 1. Uniform scale.

### Addition

The simplest nomograph is one used for addition. It is not used in practice because of its simplicity, but it illustrates well the fundamentals of nomograph construction. It has three uniform, parallel scales properly matched and spaced.

The formula of the nomograph used to add numbers is

$$L + M = N$$

and is shown in the completed form in Figure 2.

Standard graph paper, the type with major divisions 1 in. apart and ten subdivisions to the inch is an extremely useful aid and will simplify construction of a nomograph. Or, an engineer's scale, the author's choice for this work, can be used with blank paper. Using a scale has the added advantage that, where necessary, different lengths and divisions can be plotted, depending on the particular scale selected.

When constructing this type of nomograph, first draw a pair of parallel scales 5 in. apart and 4 in. long. Then draw the divisions and number the scales as shown in Figure 2.

Next, determine the location of the third scale and the range and placement of values required to satisfy the equation. For this nomograph, the location, etc., will be determined graphically. To do this, draw several lines from L to M that have the same sum; these lines will intersect on the third scale. For example:

$$\begin{aligned} L + M &= N \\ 20 + 10 &= 30 \\ 12 + 18 &= 30 \end{aligned}$$

and

$$\begin{aligned} 10 + 0 &= 10 \\ 2 + 8 &= 10 \end{aligned}$$

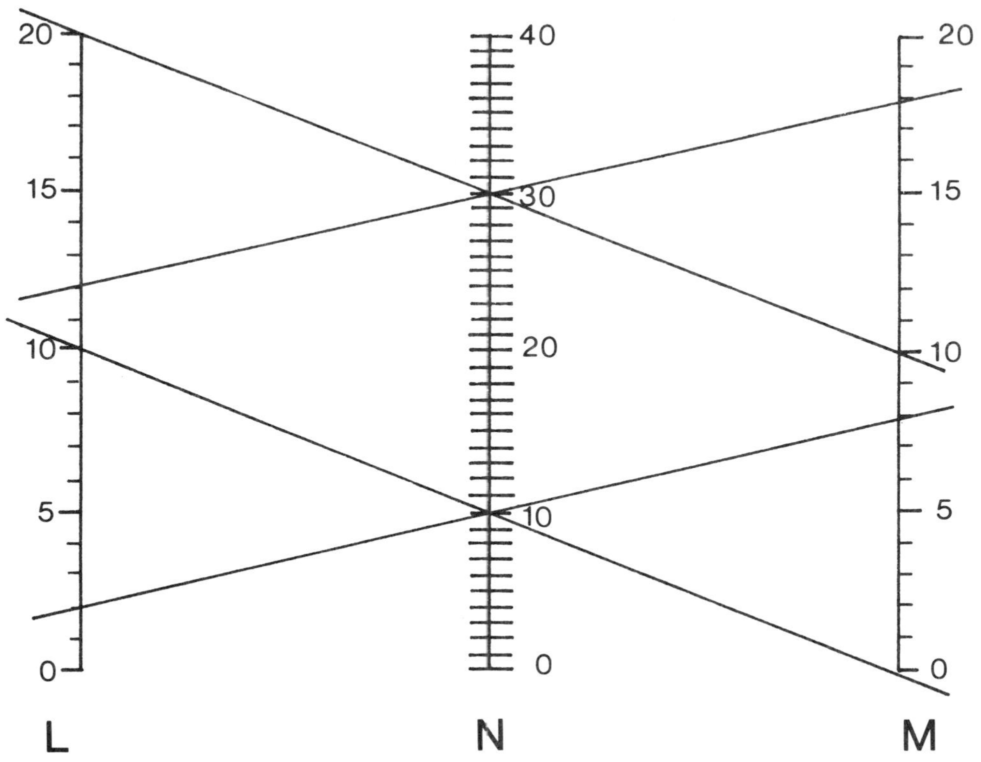


Figure 2. L + M = N.

Draw a line through the two intersecting points and you will have located scale N. Next, draw a line connecting the value 20 on the two outside scales and you will have located the range of scale N as 40. Now draw a line connecting the zero value on the two outside scales and you will have located the zero point on scale N. With the appropriate scale on the engineer's scale, draw in all the necessary divisions. Observe that it is a uniform scale because the outside scales are uniform, and it is located at the midpoint because the outside scales are identical in range.

When either L or M are scaled differently (i.e., the value of the division marks on the scales are different), the third scale shifts toward the larger-reading scale. Construct a nomograph where L is 0 to 20 and M is 0 to 40, each is 4 in. long and 5 in. apart, then locate scale N. You will find that scale N reads 0 to 60 and has moved nearer to M. The intent of this simple nomograph is to show a concept. This nomograph may also be used for subtraction:

$$N - L = M$$
$$N - M = L$$

**Logarithmic Scales**

Figure 3 shows a logarithmic scale. The spaces between the division marks are not uniform but vary according to the logarithms of the numbers on the scales.

Slide rule scales can be used to draw the logarithmic scales. Note that the slide rule scales C and D read from 1 to 10 and the scales A and B read from 1 to 10 to 100. There is no zero on log scales because log 0 = infinity.

When using a log slide rule, you multiply by adding the logarithms of the numbers. The same principle applies to the nomograph. While both the slide rule and the nomograph are scaled for the logarithms of the numbers, you see only the numbers. For example:

$$L \times M = N$$

This is equal to:

$$\log L + \log M = \log N$$

if $L = 20$ and $M = 30$,

$$N = 20 \times 30 = 600$$

and $\log 20 + \log 30 = \log N$

$$\log 20 = 1.3010$$
$$\log 30 = 1.4771$$
$$\log N = 2.7781$$

then $N = \text{antilog } 2.7781 = 600$

This is the common or Briggsian system of logarithms using log base 10. For this problem we used a four-place common log table.

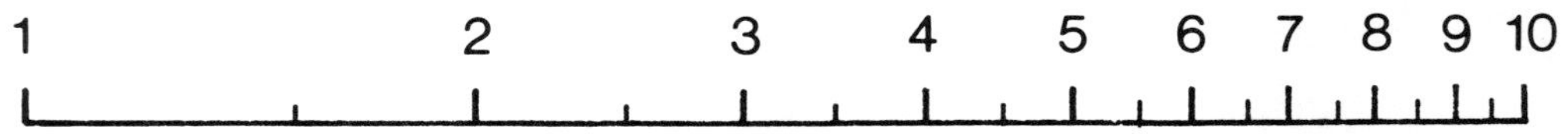


Figure 3. Logarithmic scale.

The C scale of a slide rule was used to construct the scale in Figure 3. You could also use semilog graph paper for that purpose, or you could use a four-place common log table and plot the divisions of the scale. Using a log table you will find the following plotting points:

| | |
|---|---|
| log 1 = 0.0000 | log 6 = 0.7782 |
| log 2 = 0.3010 | log 7 = 0.8451 |
| log 3 = 0.4771 | log 8 = 0.9031 |
| log 4 = 0.6021 | log 9 = 0.9542 |
| log 5 = 0.6990 | log 10 = 1.0000 |

With the use of an engineer's scale you can plot the numbers 1 through 10, and the length of the scale drawn will depend on the particular scale selected on the engineer's scale. The 10th scale will provide a log scale 10 in. long. Using the 20th scale, the log scale will be 5 in. long. The length of the scale used should satisfy the needs of the problem.

## Multiplication

We will now develop a nomograph that is used for multiplication. This type of nomograph is frequently used to solve practical engineering problems having three variables. The difference between this nomograph and the previous one is that logarithmic scales are substituted for uniform scales.

Figure 4 is a three-variable, parallel, log-scaled nomograph constructed to perform multiplication. The formula for it is:

$$L \times M = N$$

To construct this nomograph, first draw a pair of parallel log scales 5 in. apart and 5 in. long, each plotted from 1 to 10 starting with one at the bottom. You could use either the C scale of a 5-in. slide rule or plot the numbers using the log table and the 20th scale of an engineer's scale. A still simpler method is to use semilog graph paper since it has all the division marks and numbers printed on it.

Using the formula $L \times M = N$, establish several intersecting points to locate the third scale, scale N. For example:

| | $L \times M = N$ |
|---|---|
| Lines 1 & 2 | $10 \times 8 = 80$ |
| Lines 3 & 4 | $5 \times 2 = 10$ |

Now draw lines 1 & 2 and 3 & 4 as shown in Figure 4. The intersection of these lines form two points that lie on scale N. Next, draw a line through these two points that matches the lengths of scales L and M. Since these two scales are identical, you will note that scale N is located at the midpoint between the two outer scales. You should verify this by measuring the distances between the scales. Locate the top and bottom of scale N by drawing a connecting line between the tops and bottoms of scales L and M. The value of the number at the top of scale N is 100, and at the bottom is 1. You have now located the numbers 1, 10 and 100 on scale N. The distance between 1 and 10 on scale N is half the distance between 1 and 10 on scales L and M. In other words, it matches the B scale on a slide rule, so now you can use the B scale to plot the balance of the divisions and numbers on scale N. You have now constructed a nomograph that satisfies the equation $L \times M = N$ and the points that lie on a straight line cutting the three scales satisfy the equation.

In cases where the two outer scales are not identical (i.e., the value of the division marks on the scales are different) the third scale shifts toward the larger-reading scale. Construct

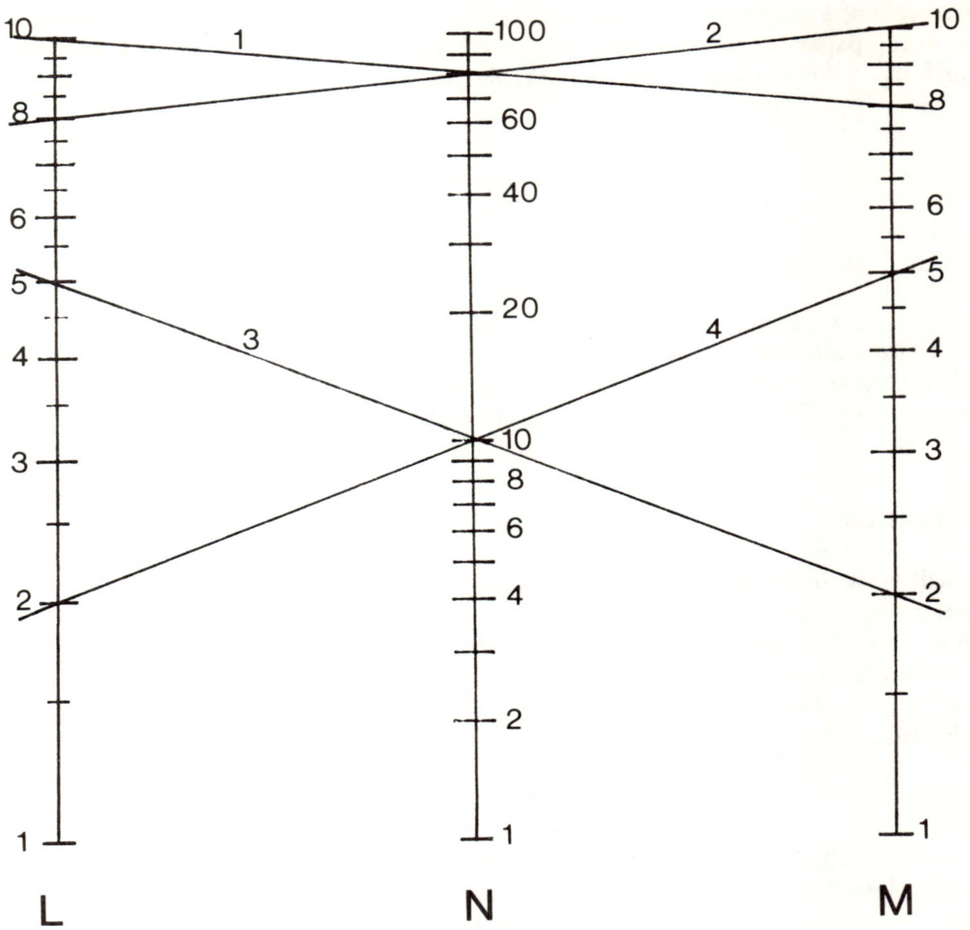


Figure 4. L × M = N.

a nomograph where L is 1 to 10 and M is 1 to 100, each is 5 in. long and 5 in. apart, then locate scale N. When this is complete, use the K scale of a slide rule to plot the divisions, and number the scale. This new scale will read from 1 to 1000.

This nomograph may also be used for division:

$$N \div M = L$$
$$N \div L = M$$

## Z-Type Nomograph

Another type of nomograph that can be used for multiplication, and easier to construct is the Z-type nomograph. It has uniform scales for the parallel scales and can be used for problems where the scales start at zero. (Remember, however, log scales cannot start at zero.) The scale connecting the parallel scales is very easily and simply constructed by plotting the values from a table.

Figure 5 shows a three-variable Z-type nomograph with two uniform parallel scales connected by a nonuniform scale. The parallel scales are plotted in opposite directions, and the third scale connects the zero points of the two parallel scales. The formula for it is:

$$L \times M = N$$

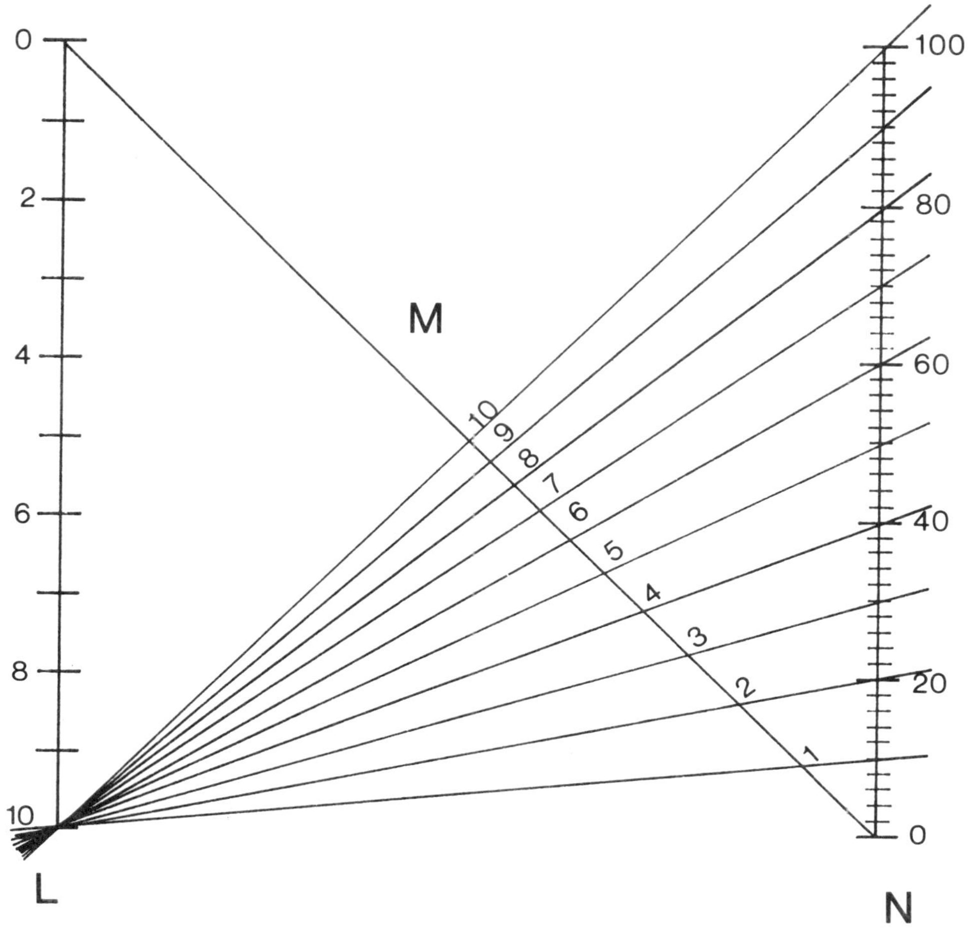


Figure 5. L × M = N.

The first step is to draw a pair of parallel scales 5 in. long and 5 in. apart. For this type of nomograph use graph paper with appropriate subdivision spacing to permit rapid and accurate drafting. For this problem where L has values from 0 to 10, we selected the 10 × 10 per inch graph paper.

Next, construct a table as shown below for all values of the equation within the range of the problem at hand. In this case both L and M range from 1 to 10.

| | |
|---|---|
| L × M = N | 10 × 6 = 60 |
| 10 × 1 = 10 | 10 × 7 = 70 |
| 10 × 2 = 20 | 10 × 8 = 80 |
| 10 × 3 = 30 | 10 × 9 = 90 |
| 10 × 4 = 40 | 10 × 10 = 100 |
| 10 × 5 = 50 | |

Using the values determined by the equation and listed in the table, complete the parallel uniform scales. Then draw in the third scale connecting the zero points of the parallel scales. Finally, starting with the number 10 on L, draw lines from it to the various numbers on N. Where these lines cross, M becomes the plotting points for the values from 1 to 10 listed in the table. After the intersecting points on M are established by drawing short perpendicular lines on M, the lines connecting L and N are erased, and the nomograph is ready to use.

As can be seen, the Z-type nomograph is far easier to construct and can be made more accurate than the log type. Additionally, it does not require a log table, semilog graph paper, slide rule scales, or engineer's scale.

The size of the nomograph depends on the size of the paper you use, and the numerical value of the scale depends on the problem. You may plot as many subdivisions on M as needed.

For practice, take a sensible heat gain formula

$$Sh = 1.087 \times Q \times T$$

where Sh = sensible heat, Btuh
Q = air volume, cfm
T = temperature rise, °F

and Sh is 0 to 20,000; Q is 0 to 500; and T is 5 to 50.

After you have completed the nomograph, check it against the one included in this book.

## Scaling Factor

While there are a number of different scales that are used to construct nomographs, the two most frequently used are the uniform scale and the logarithmic scale. The distances along a uniform scale from the zero mark are proportional to the number, while the distances along a logarithmic scale vary according to the logarithms of the numbers represented on the scales.

Figure 1 shows a uniform scale that is 4 in. long and the numerical values assigned to the scale range from 0 to 20. In that example, an engineer's scale was used to plot the divisions of the scale, letting one inch equal the number five, but this can also be done mathematically. This method is particularly important when the values cannot be simply plotted with the aid of an ordinary ruler or engineer's scale.

An equation could also have been used to solve the problem. Since there are 20 divisions in a 4-in. scale, there must be some factor by which we can multiply the number on the scale to locate its position with respect to the zero mark. We will call that the scaling factor (SF) and it can be determined by the formula

$$SF = \text{length of scale/numerical value of scale}$$

In Figure 1

$$\text{length of scale} = 4 \text{ in.}$$
$$\text{numerical value of scale} = 20 - 0 = 20$$

Therefore, the scaling factor for the uniform scale is:

$$SF = 4/20 = 0.2$$

The distance from the zero mark to any point along the scale is equal to the scaling factor multiplied by the numerical value of the division mark. ($X = SF \times \text{number}$) Locate the number 15 on the scale:

$$X = 0.2 \times 15 = 3 \text{ in. from zero}$$

This scaling factor can be used to construct a three-scale parallel nomograph. To construct the nomograph shown in Figure 2, we need to know the location of the third scale and its numerical value. Earlier we plotted this scale graphically; now we will show that it can also be plotted and located mathematically.

let

$$Y = \text{distance between L and M}$$
$$a = \text{SF of L}$$
$$b = \text{SF of M}$$
$$Z = \text{distance of N from M}$$

The distance of N from M is determined by the formula:

$$X = Y \times b/(a + b)$$
$$X = 5 \times 0.2/(0.2 + 0.2) = 2.5 \text{ in.}$$

This agrees with the graphical location of N determined earlier. To complete the nomograph, we need to plot the numerical value of scale N. Let X equal the distance from the zero mark of the scale, then the value at that point is

$$N_v = X \times (a + b)/(a \times b)$$

at the top of the scale $x = 4$ in.

$$N_v = 4 \times (0.2 + 0.2)/(0.2 \times 0.2) = 40$$

Now we can construct the nomograph where L is 0 to 20, and M is 0 to 40.

$$\text{SF of L} = 4/20 = 0.2$$
$$\text{SF of M} = 4/40 = 0.1$$
$$Z = 5 \times 0.1/(0.2 + 0.1) = 1.66 \text{ in.}$$
$$N_v = 4 \times (0.2 + 0.1)/(0.2 \times 0.1) = 60$$

As can be seen, the results are exactly the same for both methods.

The scaling factors can be determined for log scales just as easily, except logarithms of the numbers are used.

$$\text{SF} = \text{length of scale/logarithm of the number}$$

The scale shown in Figure 3 is 5 in. long, and the numerical values range from 1 to 10, therefore

$$\text{SF} = 5/(\log 10 - \log 1) = 5$$

Determine the location of the number 8 on that log scale:

$$X = \text{SF} \times \log 8$$
$$5 \times 0.9031 = 4.5 \text{ in.}$$

Determine the scaling factor of the N scale of the nomograph in Figure 4.

$$\text{SF} = 5/(\log 100 - \log 1) = 2.5$$

Locate the number 8 on the N scale

$$X = 2.5 \times 0.9031 = 2.26 \text{ in.}$$

To find the distance of N from M, use the same formula used in the previous nomograph $L + M = N$. First calculate the scaling factors for L and M.

SF = length of scale/logarithm of the number
length of scale = 5 in.
logarithm of the number = log 10 − log 1

$$SF = 5/(\log 10 - \log 1)$$
$$= 5/(1 - 0) = 5$$
$$a = SF \text{ of } L = 5$$
$$b = SF \text{ of } M = 5$$
$$Z + 5 \times 5/(5 + 5) = 2.5 \text{ in.}$$

Now construct the nomograph where L is 1 to 10, and M is 1 to 100.

$$a = SF \text{ of } L = 5/(\log 10 - \log 1) = 5$$
$$b = SF \text{ of } M = 5(\log 100 - \log 1) = 2.5$$
$$Z = 5 \times 2.5/(5 + 2.5) = 1.666$$

Again, the results are the same with both methods.

Before the third scale of a three-variable nomograph can be plotted, you must locate one point on each scale that satisfies the equation. This provides a starting point on the N scale. For example, for the $L + M = N$ nomograph, both the zero point and $20 + 20 = 40$ point must be located. For the $L \times M = N$ nomograph, the base and top, or any other point, may be used. If the base and top points are located, then the scale can be plotted from 1 to 100.

Construct a nomograph to determine for the equation

$$kW = cfm \times \Delta T/3193$$

Where the scales have the following ranges

- $N = kW = 1$ to 150
- $M = cfm = 100$ to 10,000
- $L = \Delta T = 1$ to 100

and L and M are 9 in. long and 5 in. apart.

For L,

$$SF = 9/(\log 100 - \log 1)$$
$$= 9/(2 - 0) = 4.5$$

For M,

$$SF = 9/(\log 10{,}000 - \log 100)$$
$$= 9/(4 - 2) = 4.5$$
$$Z = 5 \times 4.5/(4.5 + 4.5) = 2.5 \text{ in.}$$

Next locate two points on the N scale, for example 1 and 100.

$$\Delta T = 3193 \times kW/cfm$$

where

$$kW = 1$$
$$cfm = 1000$$
$$\Delta T = 3193 \times 1/1000 = 3.193$$

and

$$kW = 100$$
$$cfm = 10{,}000$$
$$\Delta T = 3193 \times 100/10{,}000 = 31.93$$

The measured distance on scale N between kW = 1 and 100 is now 4.5 in. Now determine the scaling factor for N and then plot it.

$$SF = 4.5/(\log 100 - \log 1) = 2.25$$

We have shown that the N scale is located midway between L and M and starts from the line from 1000 cfm through 1 kW to 3.193 F. All that is left to do is plot the scale. Locate the point 150 kW with respect to 1 kW.

$$X = SF \times \log 150$$
$$2.25 \times 2.1761 = 4.9 \text{ in.}$$

## HOW TO USE NOMOGRAPHS

Because this book is a collection of nomographs, it is important to explain to readers how to use a nomograph. In effect we have explained how to use them above when we say to align one scale with another and read the answer where the straight line cuts the third scale. However, with additional information, we hope you will be able to make maximum use of nomographs.

As an example, in a water flow problem, the quantity and velocity of the water and the pipe size is determined from the formula Q = AV. To solve this problem mathematically, you plug in the known parameters and solve the problem. (Providing, of course, that you know the cross-sectional area of the pipe; if not, you must either solve for it or look it up in a handbook.) For another set of conditions, you do more calculations, and with every calculation, there is the ever-present possibility of error. Using a nomograph, that possibility has been eliminated, and you do not even have to know the area of the pipe, because the nomograph was designed for pipe diameters. In actual practice the only pipe dimension you need is diameter.

Suppose that in a water flow problem, the quantity of water is fixed and you want to limit the velocity to some maximum amount, but the pipe diameter is determined to be 7 in. Since this size is not commercially available, you merely shift the straightedge until you obtain the parameters that provide the optimum solution to your problem. This technique permits you to make better use of nomographs.

Second, extending (or reducing) the scales to expand the use of nomographs is another method to enhance their use. To do this successfully, however, you need to understand the formula from which the nomograph was constructed.

Take, for example, the water flow problem Q = AV. If you double the range of the quantity scale (2Q), you must double the velocity scale (2V), or 2Q = A(2V). Any factor not equal to 1 will work in this formula. However, you cannot double or halve the pipe diameter in this problem and still use the nomograph. Compare the area of a 4-in. pipe to that of a 2-in. pipe and see what happens.

Unfortunately, not all nomographs are so designed that their scales can be extended. Fortunately, however, many nomographs are designed to solve specific problems and there is no need to extend the scales. Always carefully read what authors write in the text for their nomographs. If you decide to experiment with extending scales, insert your extended (or reduced) numbers into the formula and check it against the solution obtained from the nomograph.

Understanding the accuracy of nomographs can be an additional asset in the use of nomographs. The degree of accuracy of nomographs is greatest when the straightedge is most nearly perpendicular to the scales, and the accuracy is the least when the straightedge approaches parallel to any one or more scales. Further, the degree of accuracy depends on the ability to align the straightedge with the divisions of the scales and the interpolation when the straightedge falls between the divisions of the scales.

You may use a nomograph to zero in on the optimum solution to the problem quickly. You may also change the parameters and obtain additional answers quickly and easily.

## REFERENCE

1. *McGraw-Hill Dictionary of Scientific and Technical Terms* (New York: McGraw-Hill, 1974).
t/nt

# GROUP 1
# AIR

## 1–1. Converts acfm at elevated temperatures to scfm

The nomograph on the following page is useful in determining the volumetric flowrate of air at standard conditions, scfm, when the actual flowrate has been measured at elevated temperatures, acfm.

The flowrate at standard conditions is obtained from the formula:

$$\text{scfm} = \text{acfm} \times 530/(460 + T)$$

where $T$ = dry bulb temperature of measured air stream, °F

**EXAMPLE**

Determine the scfm of an air stream measured at 3000 cfm and 250°F.

**Solution**

Extend a line from 3000 on scale A through 250 on scale T to scale S. There, read a value of 2240 scfm.

Note that for air quantities beyond range of scales A and S, the scale readings in the solution must be multiplied by the same factors used on the known scale. For example, suppose the measured cfm in the above problem were 6000 cfm. We would multiply the solution by 2, the same number used to multiply 3000 by to obtain 6000. The result would be $2240 \times 2 = 4480$ scfm.

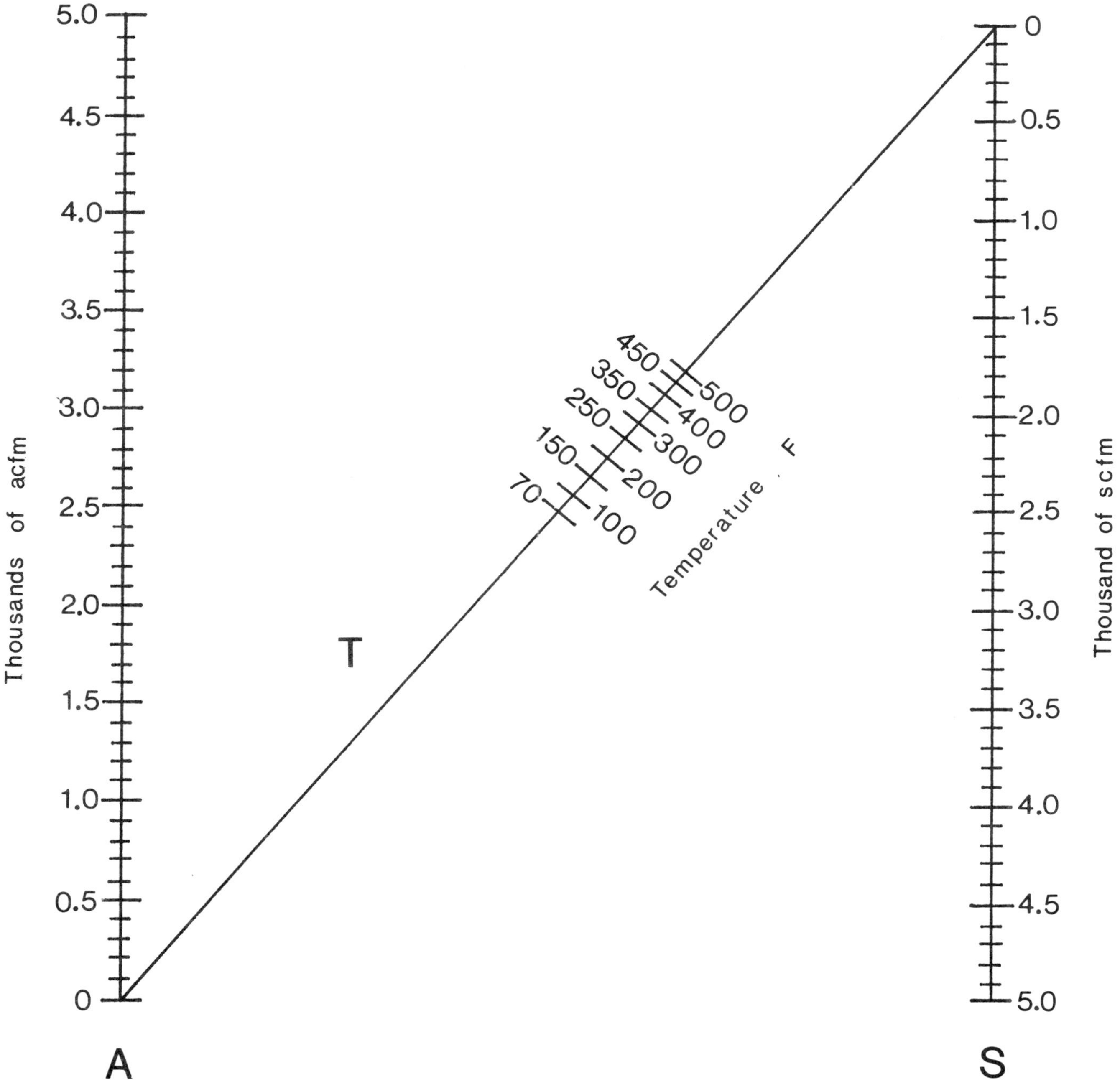
5.0
4.5
4.0
3.5
3.0
2.5
2.0
1.5
1.0
0.5
0
Thousands of acfm
A
T
70
100
150
200
250
300
350
400
450
500
Temperature F
0
0.5
1.0
1.5
2.0
2.5
3.0
3.5
4.0
4.5
5.0
Thousand of scfm
S

## 1–2. Properties of moist air

This nomograph permits the quick calculation of dewpoint temperatures for combination of temperature and relative humidity. Simply laying a straightedge across from a given air temperature to a given relative humidity will permit the reading of the dewpoint temperature. The same procedure may be used to determine either of the other factors if the dewpoint and the second factor are known.

### EXAMPLE

Determine the dewpoint temperature when the dry bulb (db) temperature is 70°F and the relative humidity (rh) is 70%.

### Solution

Extend a line from 70°F on scale A to 70% on scale C and read 60°F dewpoint temperature on scale B.

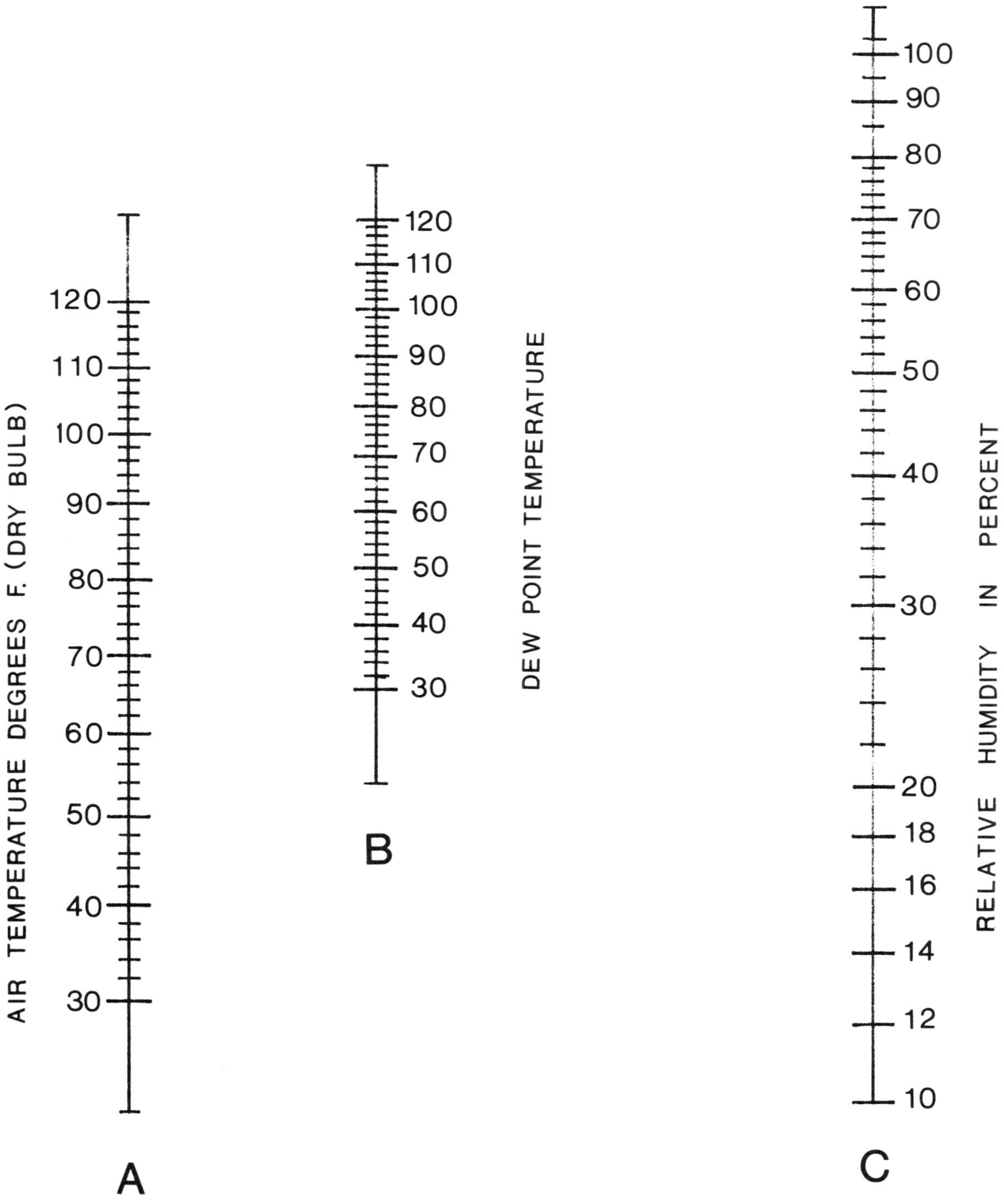
AIR TEMPERATURE DEGREES F. (DRY BULB)
120
110
100
90
80
70
60
50
40
30
A
120
110
100
90
80
70
60
50
40
30
DEW POINT TEMPERATURE
B
100
90
80
70
60
50
40
30
20
18
16
14
12
10
RELATIVE HUMIDITY IN PERCENT
C

## 1-3. Determines moisture addition, removal quantities

Most psychrometric charts list moisture content of air in grains of moisture per pound of dry air. When the amount of moisture to be added or removed for a specific quantity of air is required, conversion factors must be employed to arrive at a value with familiar units. To obtain a value of water quantity in gph, the following conversion must be made. Using standard conditions:

$$W = Q \times 0.075 \times 60 \times M \times 1/7000 \times 1/8.337$$

where

| | | |
|---|---|---|
| W | = | amount of moisture added or removed, gph |
| Q | = | quantity of air, cfm |
| M | = | change in moisture content of air, gr/lb dry air |
| 0.075 | = | density of standard air, lb/ft$^3$ |
| 60 | = | min/hr |
| 1/7000 | = | lb water/gr |
| 1/8.337 | = | gal/lb water |

Simplifying:

$$W = 7.71089 \times 10^{-5}\ Q\ M$$

If the moisture content is given in lb water/lb dry air rather than in gr/lb dry air, then a different overall conversion factor is derived. By using the nomograph, the conversion factors are already incorporated, making a designer's life somewhat easier. Selecting humidifier capacity for space requirements or sizing a drain line for condensate removal from cooling coils can be done quickly by using the nomograph on the following page.

### EXAMPLE

What must the capacity of a humidifier be to bring 6000 cfm of outside air at 40°F db and 81% rh up to a room condition of 76°F db and 52% rh?

### Solution

From the psychrometric chart, the change in moisture level is 40 gr/lb dry air. Extend a line from 6000 on scale Q through 40 on scale M, and read the answer on scale W as 18.5 gph of water or 154 lb/hr of steam.

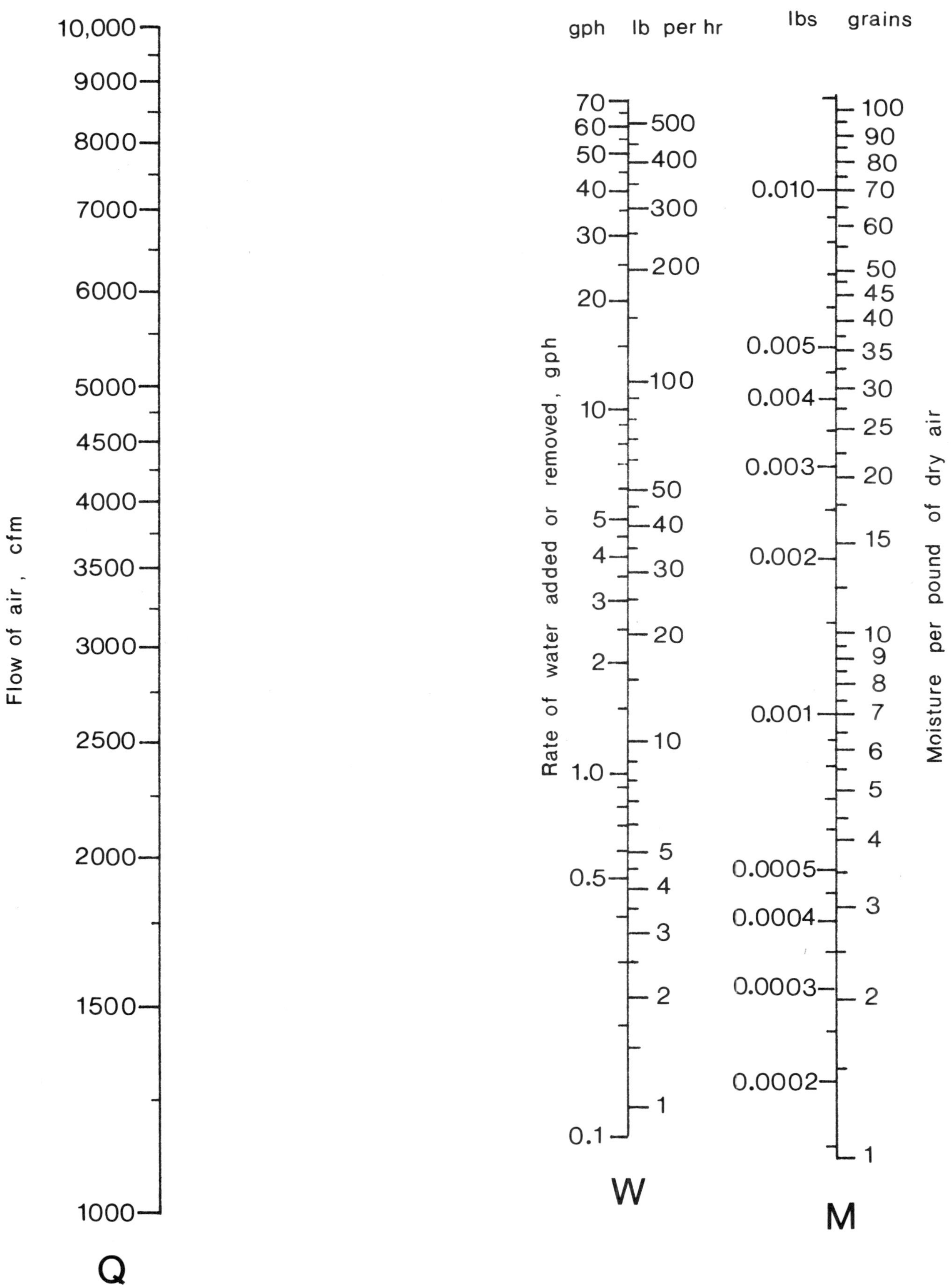
Flow of air, cfm
10,000
9000
8000
7000
6000
5000
4500
4000
3500
3000
2500
2000
1500
1000
Q
gph
lb per hr
Rate of water added or removed, gph
70
60
50
40
30
20
10
5
4
3
2
1.0
0.5
0.1
500
400
300
200
100
50
40
30
20
10
5
4
3
2
1
W
lbs
grains
Moisture per pound of dry air
0.010
0.005
0.004
0.003
0.002
0.001
0.0005
0.0004
0.0003
0.0002
100
90
80
70
60
50
45
40
35
30
25
20
15
10
9
8
7
6
5
4
3
2
1
M

## **1-4.** Calculates temperature of mixed air

A common problem in ventilation cooling and air conditioning is determining the final air temperature of a mixture of two different air quantities at two different initial temperatures. This nomograph will determine quickly and accurately the final air temperature for a mixture of any two quantities of air when the initial dry bulb temperature and the percentage of the final air quantity are known. In air conditioning problems, the temperature of the air entering the cooling coil can be determined; in ventilation cooling, the supply air temperature can be determined.

### EXAMPLE

Determine the final temperature of the mixed air when 25% outside air at 60°F db is mixed with 75% return air at 80°F db.

### Solution

Align 60°F on the initial temperature scale $T_i$ with 25% on the percentage scale P, and extend this line to reference line 1 (step 1). From reference line 1, project a line to reference line 2, parallel to the base line (step 2). Then align 80°F on scale $T_i$ with 75% on scale P, and extend a line to reference line 1 (step 3). For the final mixed air temperature, align this second point on the reference line 1 with the point on reference line 2 obtained in step 2, and read the answer on scale $T_f$ as 75°F db.

This nomograph can be used when any two of the three variables are known to determine the third.

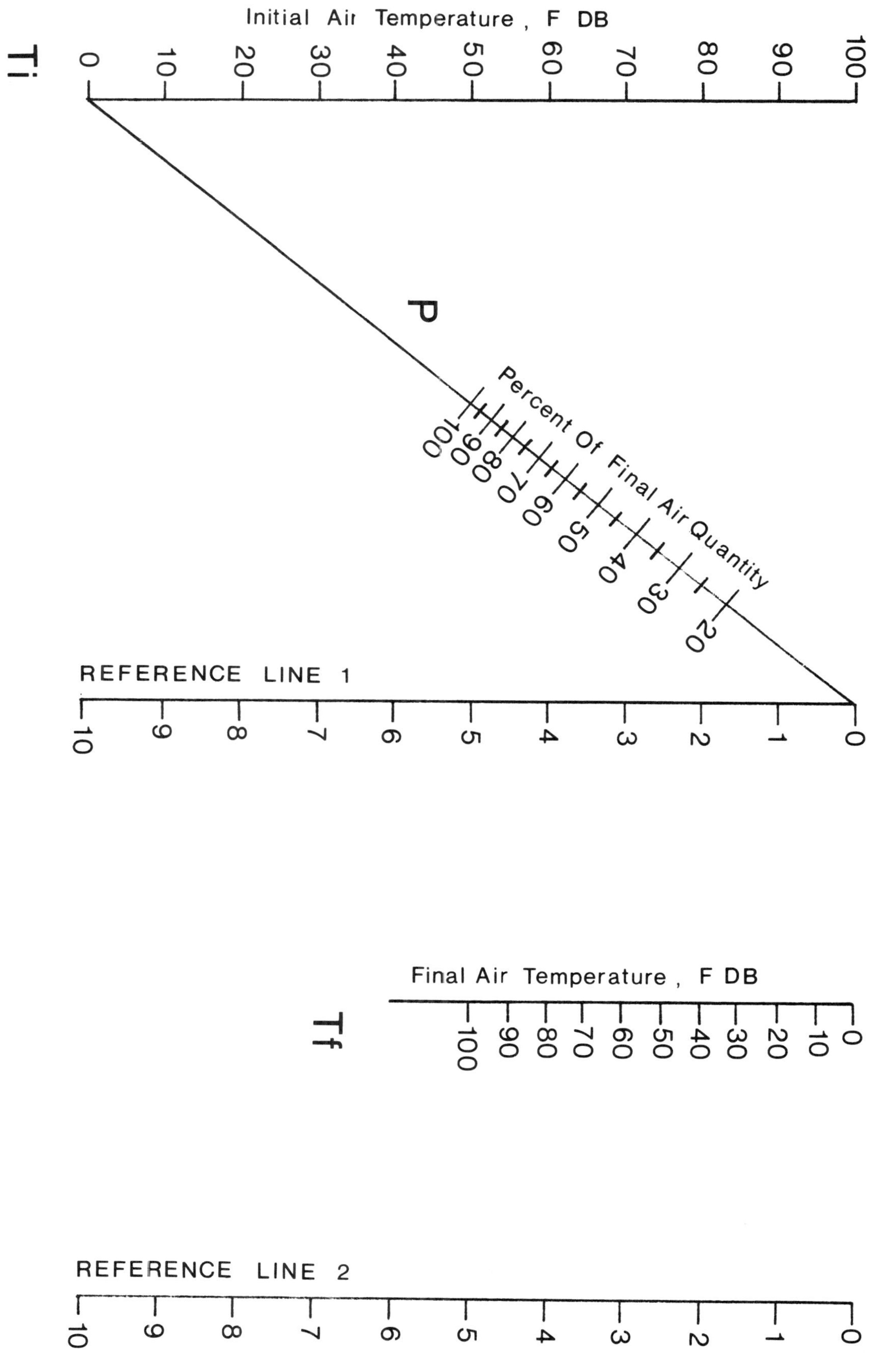
Initial Air Temperature , F DB
0
10
20
30
40
50
60
70
80
90
100
Ti
P
Percent Of Final Air Quantity
20
30
40
50
60
70
80
90
100
REFERENCE LINE 1
0
1
2
3
4
5
6
7
8
9
10
Final Air Temperature , F DB
0
10
20
30
40
50
60
70
80
90
100
Tf
REFERENCE LINE 2
0
1
2
3
4
5
6
7
8
9
10

## 1-5. Calculates required ventilation air changes

Local, state and national codes usually specify ventilation requirements for special areas in air changes per hour. The nomograph on the following page quickly and accurately determines the volume of air in cubic feet per minute that is needed to provide the required air changes. In cases where the quantity of air is known, the number of air changes produced can be determined.

### EXAMPLE

A storage room is 12 ft wide and 20 ft long with a 10-ft ceiling. The floor area is 240 $ft^2$. The ventilation requirement for this specific space use is 12 air changes per hour. Find the cfm required to satisfy this condition.

### Solution

Align 240 $ft^2$ on scale A with 10 ft on scale H and note 2400 $ft^3$ on the pivot line. Then align 2400 with 12 air changes on scale C and read the answer as 480 cfm on scale V.

With a little experimentation, this nomograph can be used to solve problems with different unknown quantities; for example, do the above problem backward.

Further, problems with floor areas beyond the range of scale A can be solved just as easily. For example, if the floor area is 960 $ft^2$ (chosen for convenience), the ceiling is 10 ft high, and 12 air changes per hour are required, what is the necessary cfm? Divide 960 $ft^2$ by some number that will produce an area whose number is near the midpoint of the scale. The number chosen was 4, and the area is now 240. Solve as before and multiply the answer, 480 by 4 to obtain the answer for this problem as 1920 cfm.

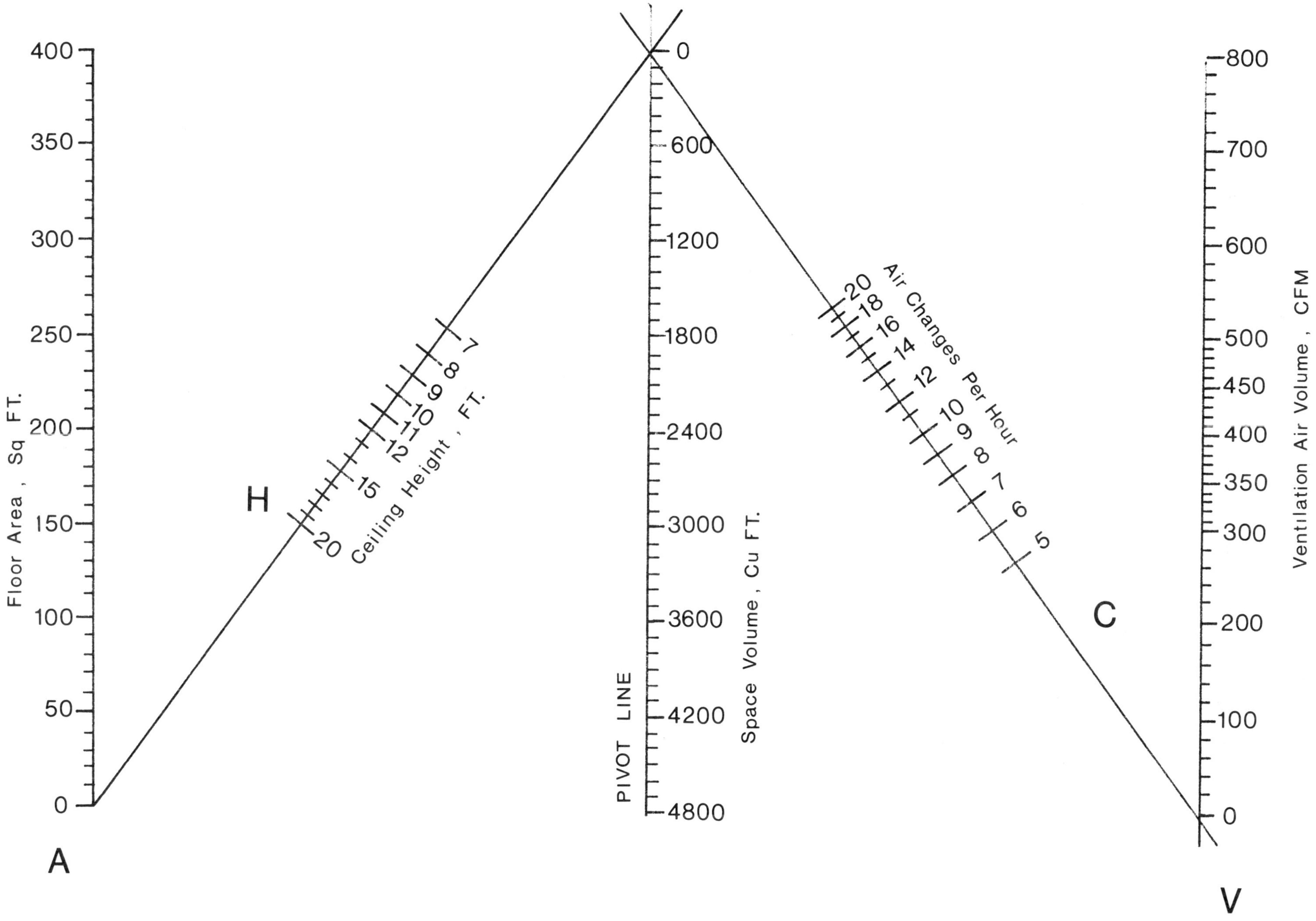
Floor Area , Sq FT.
400
350
300
250
200
150
100
50
0
A
H
Ceiling Height , FT.
7
8
9
10
11
12
15
20
PIVOT LINE
Space Volume , Cu FT.
0
600
1200
1800
2400
3000
3600
4200
4800
Air Changes Per Hour
20
18
16
14
12
10
9
8
7
6
5
C
Ventilation Air Volume , CFM
800
700
600
500
450
400
350
300
200
100
0
V

## 1-6. Determining air requirements of exhaust hoods—slot velocity

Many plants have processes that generate fumes, gases, vapors, mists or particulates. Industrial ventilation systems designed to capture contaminants at the source are the most effective method of preventing their spread through the plant.

The basic equation for ventilation calculations is:

$$Q = A \times V$$

where
$Q$ = air volume, cfm
$A$ = cross-sectional area through which air flows, $ft^2$
$V$ = air velocity, fpm

This nomograph can be used to find exhaust-hood air requirements if two of the three variables are known. A quick examination of the hood opening size determines whether to use nomograph 1-6 or 1-7 initially. Nomograph 1-6 is used for slot hoods, and nomograph 1-7 is for all other types. It may be necessary to use both nomographs in some calculations, depending on the conditions encountered.

The hood's outlet duct size can be determined when any two of the three variables are known.

### EXAMPLE

A 48-in.-wide hood needs a slot velocity of 150 fpm with a 200-cfm volume. What size of slot opening is required?

### Solution

On nomograph 1-6, connect 200 cfm on scale Q with 150 fpm on scale V and extend the line to scale D. Connect this point with 48 in. on scale L, and where scale W is crossed read the required slot width as 4 in.

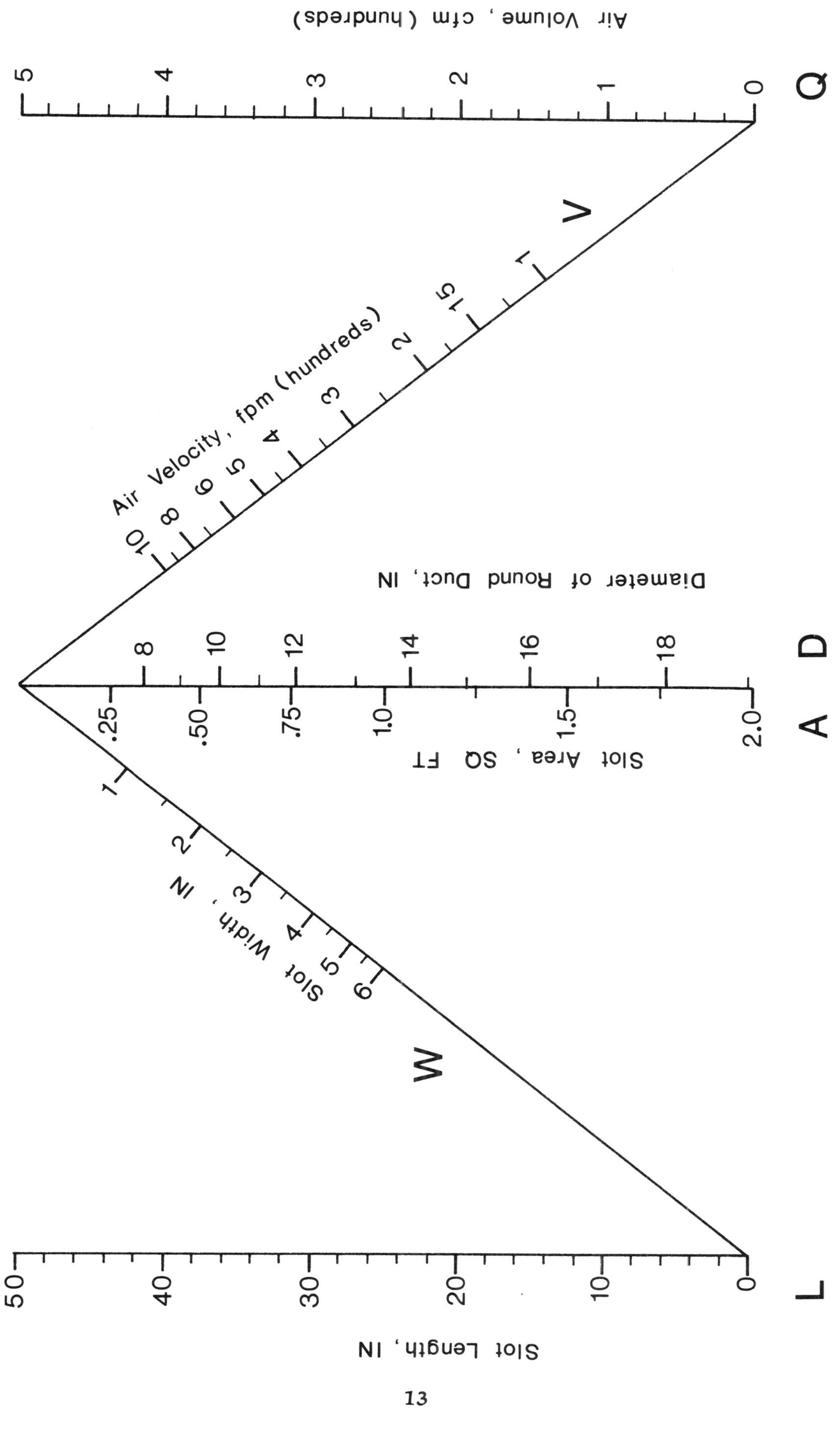
Air Volume, cfm (hundreds)
Q
V
Air Velocity, fpm (hundreds)
Diameter of Round Duct, IN
Slot Area, SQ FT
A
D
Slot Width, IN
W
Slot Length, IN
L

## 1-7. Determining air requirements of exhaust hoods—face velocity

Many plants have processes that generate fumes, gases, vapors, mists or particulates. Industrial ventilation systems designed to capture contaminants at the source are the most effective method of preventing their spread through the plant.

The basic equation for ventilation calculations is:

$$Q = A \times V$$

where $Q$ = air volume, cfm
$A$ = cross-sectional area through which air flows, $ft^2$
$V$ = air velocity, fpm

This nomograph can be used to find exhaust-hood air requirements if two of the three variables are known. A quick examination of the hood opening size determines whether to use nomograph 1-6 or 1-7 initially. Nomograph 1-6 is used for slot hoods, and nomograph 1-7 is for all other types. It may be necessary to use both nomographs in some calculations, depending on the conditions encountered.

The hood's outlet duct size can be determined when any two of the three variables are known.

### EXAMPLE

A hood has a face opening 23 in. long and 17 in. wide, with a face velocity of 150 fpm. What is the required air volume?

### Solution

On nomograph 1-7, connect 23 in. on scale L with 17 in. on scale W and extend the line to scale A. Connect this point with 150 fpm on scale V and extend the line to scale Q, there read the required air volume as 410 cfm.

If a 1000-fpm transport velocity is required to move the particulates in the air stream, what size round duct is needed on the hood outlet? An examination of nomograph 1-7 reveals that a combination of 410 cfm and 1000 fpm does not exist on the companion scales, so nomograph 1-6 must be used.

On nomograph 1-6, connect 410 cfm on scale Q with 1000 fpm on scale V, extend the line to scale D, and read the required round duct diameter of 8-9 in. The smaller 8-in. size should be used to meet the minimum air-velocity requirement.

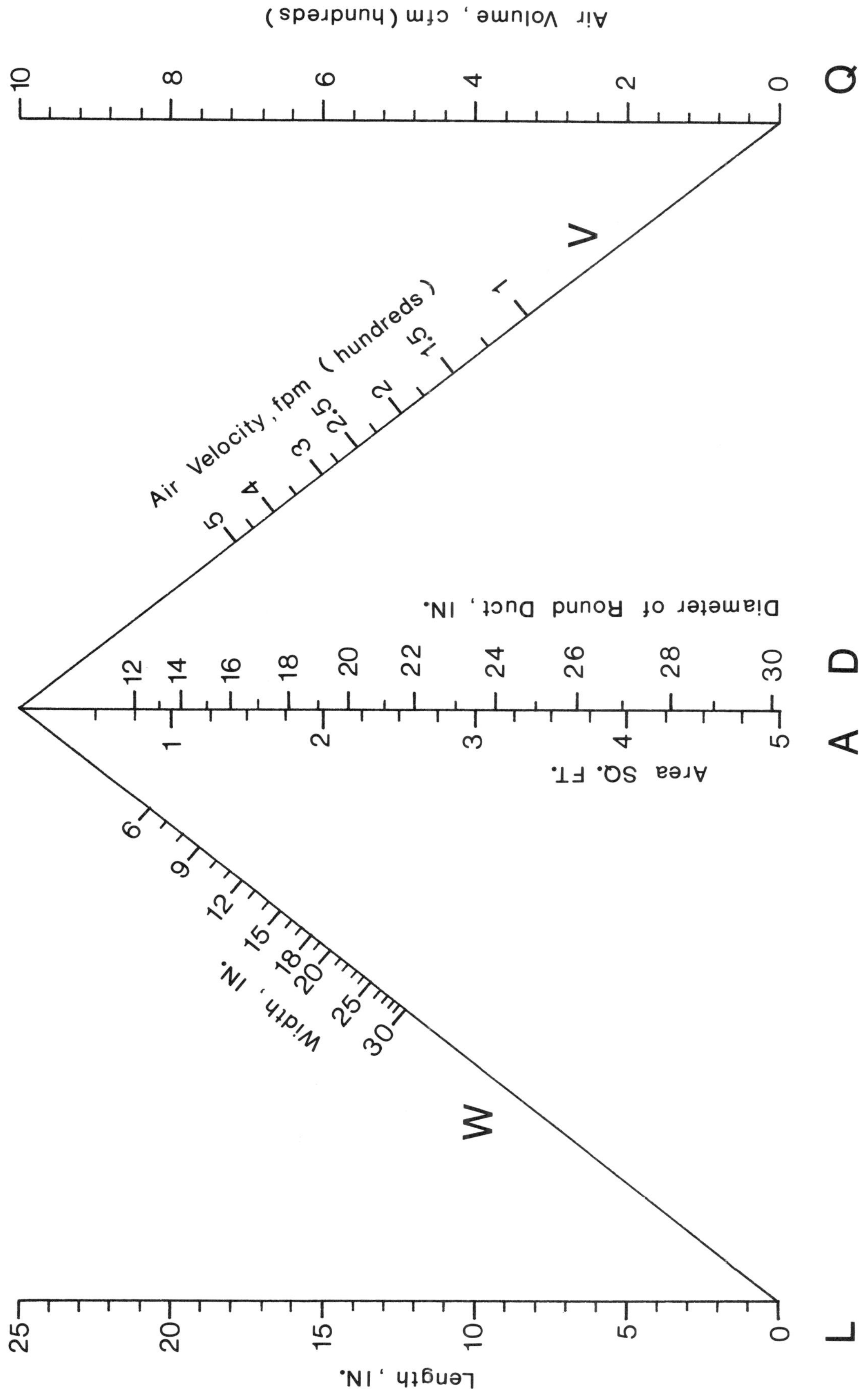
Air Volume, cfm (hundreds)
Q
Air Velocity, fpm (hundreds)
V
Diameter of Round Duct, IN.
D
Area SQ. FT.
A
Width, IN.
W
Length, IN.
L

## 1-8. Transport air-velocity calculator

Building codes (NFPA, OSHA, etc.) establish the transport velocities required to move particulates in the air stream in exhaust systems and for pneumatic conveying. These requirements are covered in NFPA No. 91, Blower and Exhaust Systems, Industrial Ventilation by American Conference of Governmental Industrial Hygenists, and OSHA. Transport velocity is defined as the minimum air velocity necessary to move particulate matter in the air stream and is obtained from the formula:

$$V = Q/A$$

where $V$ = average linear velocity, fpm
$Q$ = quantity of air flowing, cfm
$A$ = cross-sectional area of duct, $ft^2$

### EXAMPLE

What quantity of air is required in a 10-in. duct when 3000 fpm is the transport velocity?

### Solution

Align 3000 fpm on scale A with 10 in. duct on scale C and read 1650 cfm on scale E.

### EXAMPLE

The transport velocity is 3000 fpm and the duct is 3 in. What air quantity is needed?

### Solution

Align 3000 fpm on scale A with 3 in. on scale B and read 150 cfm on scale D. Note: Scales B and D, and C and E are companion scales. Scales B and D are provided to read more accurately when smaller size ducts are used.

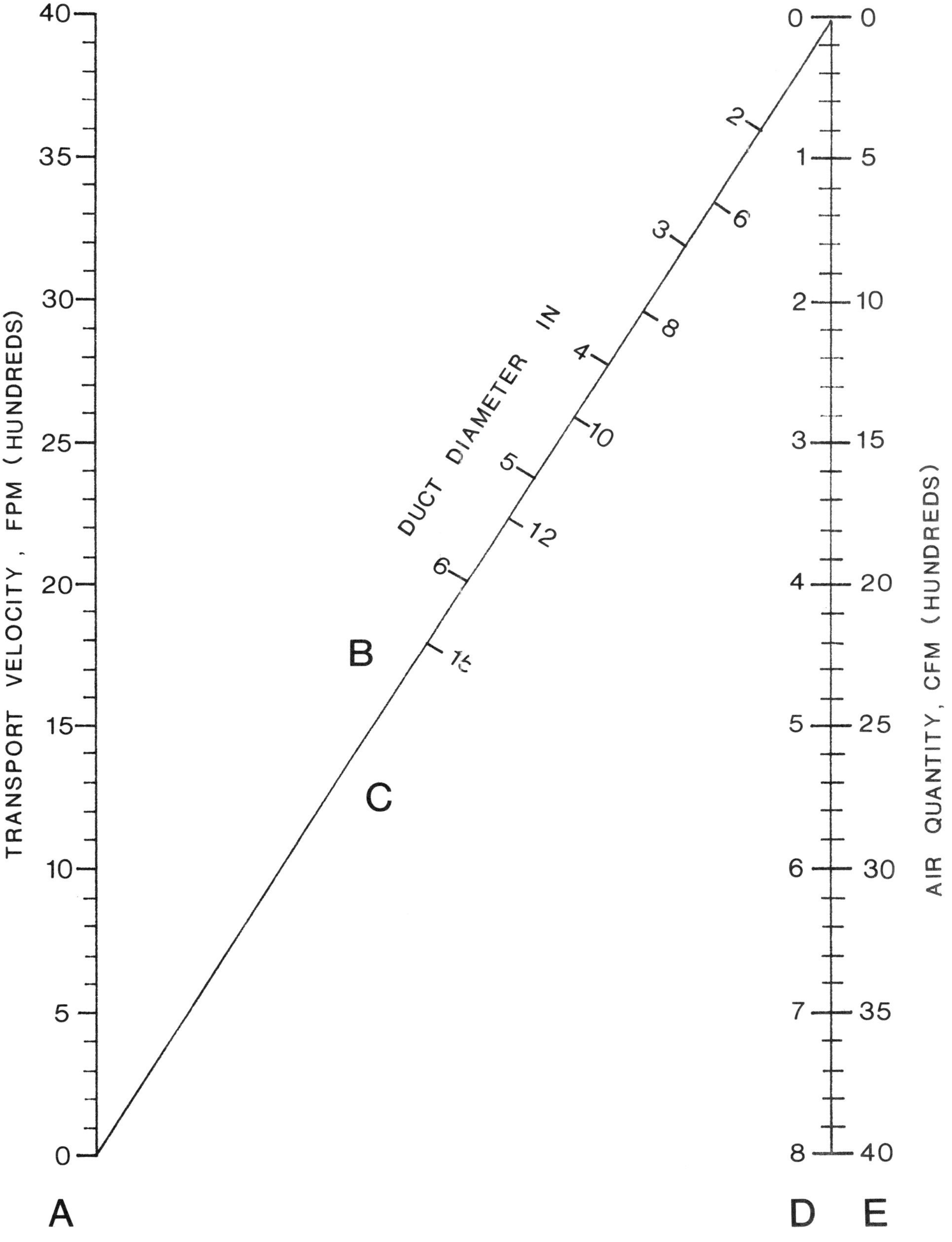
TRANSPORT VELOCITY, FPM (HUNDREDS)
0
5
10
15
20
25
30
35
40
DUCT DIAMETER IN
2
3
4
5
6
6
8
10
12
15
B
C
AIR QUANTITY, CFM (HUNDREDS)
0
1
2
3
4
5
6
7
8
0
5
10
15
20
25
30
35
40
A
D
E

## 1-9. Friction loss chart

With this nomograph you can quickly and easily determine the friction loss in round ducts frequently used in exhaust systems when any three of the four parameters are known.

### EXAMPLE

Determine the friction loss in inches of water in a 6.5-in.-diameter pipe 53 ft long with an air velocity of 4250 fpm.

### Solution

Align 6.5 in. on scale A with 53 ft on scale B and mark the intersection on line R. Align this mark with 4250 fpm on scale C and read 2.2 in. loss on scale D.

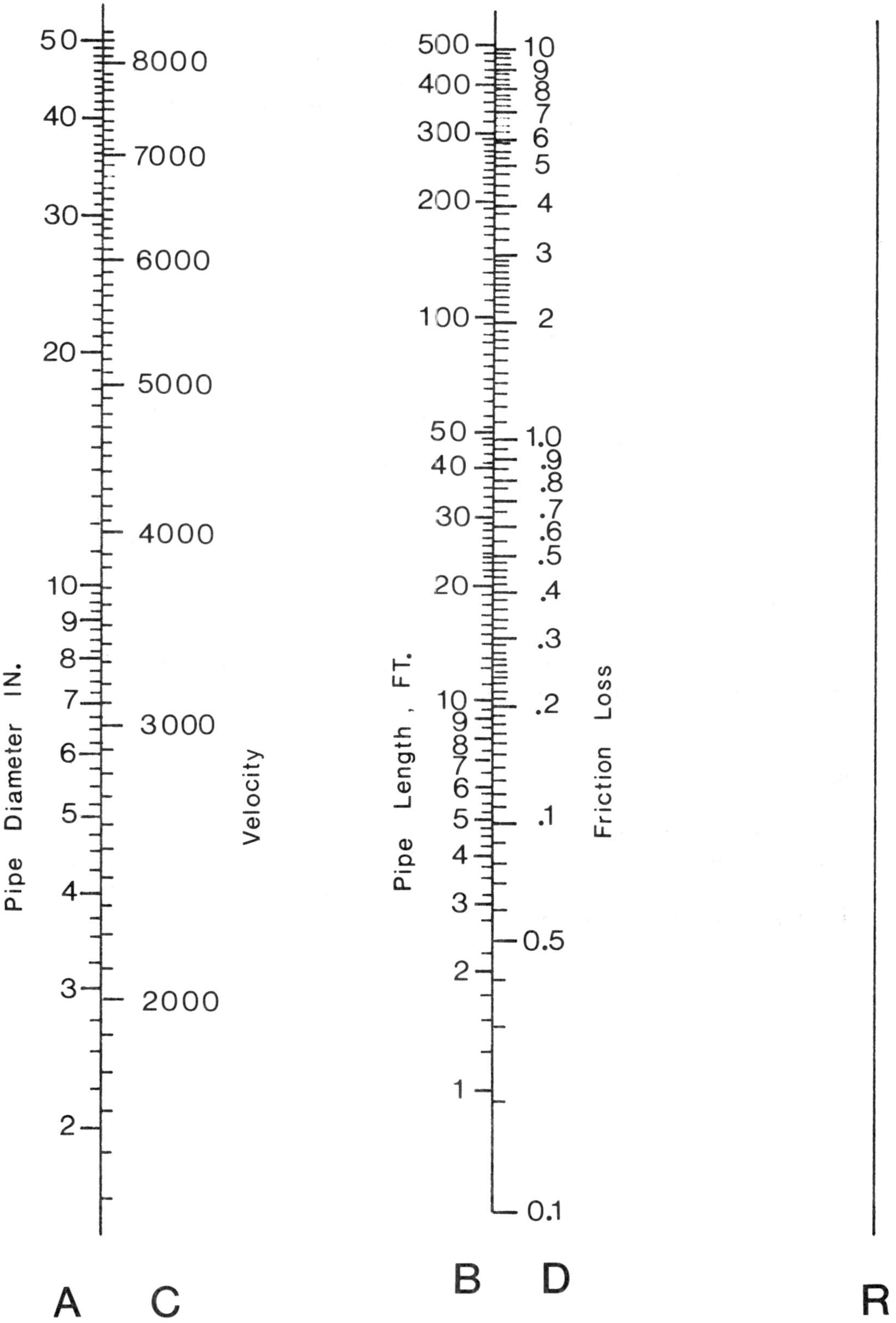
Pipe Diameter IN.
50
40
30
20
10
9
8
7
6
5
4
3
2
8000
7000
6000
5000
4000
3000
2000
Velocity
Pipe Length , FT.
500
400
300
200
100
50
40
30
20
10
9
8
7
6
5
4
3
2
1
10
9
8
7
6
5
4
3
2
1.0
.9
.8
.7
.6
.5
.4
.3
.2
.1
0.5
0.1
Friction Loss
A
C
B
D
R

# 1-10. Determines parameters for HVAC electric duct reheating

This simple nomograph makes it possible to quickly and easily solve HVAC electric reheat problems, or for that matter, any problem where electric heating coils are used to heat air passing over the coils for primary heating, supplementing heat pumps, auxiliary heat, air conditioning reheating.

Sensible heat gain is obtained by the formula:

$$Sh = 1.087 \times Q \times T$$

where Sh = sensible heat, Btuh
Q = air volume, cfm
T = temperature rise, °F

In a typical electric heating problem, four parameters are necessary to properly specify and schedule the heating unit. These are: sensible heat, temperature rise, air volume and electrical power. Therefore, with this nomograph and a straightedge, you can quickly determine the unknown parameter.

## EXAMPLE

Determine the temperature rise of air and the kilowatt rating of the heating coil when Sh is 8 MBtuh and Q equals 350 cfm.

### Solution

Align 350 cfm on the Q scale with 8 MBtuh on the Sh scale, and at the point of intersection on the T scale read 21°F. Also, directly opposite on the Sh scale read the size of the electric reheat needed on the kW scale. In this case, the answer is 2.35 kW.

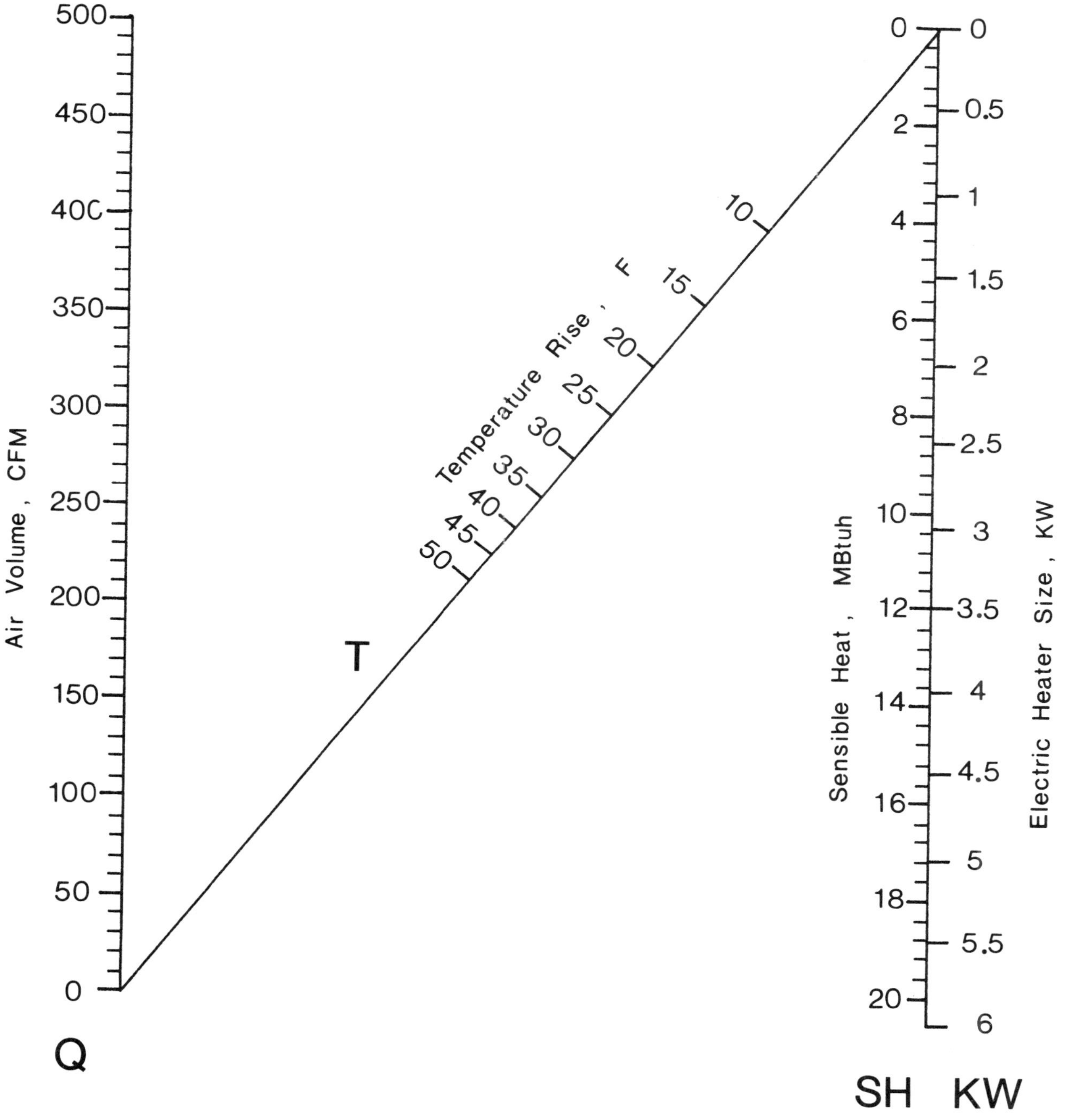

Air Volume , CFM
500
450
400
350
300
250
200
150
100
50
0
Q
T
Temperature Rise , F
10
15
20
25
30
35
40
45
50
Sensible Heat , MBtuh
0
2
4
6
8
10
12
14
16
18
20
SH
Electric Heater Size , KW
0
0.5
1
1.5
2
2.5
3
3.5
4
4.5
5
5.5
6
KW

## 1-11. Calculating the capacity of electric heating coils

Electric heaters are often used in industrial plants for space and makeup air heating, and for duct reheating in air-conditioning systems. This nomograph can be used to calculate required heating coil capacity to raise the temperature a predetermined amount for a given quantity of air. The current that will flow through the circuit for all standard voltages on single and three-phase systems can also be determined from the nomograph.

The nomograph is based on air with a specific heat of 0.237 Btu/lb/°F and a density of 0.08 $lb/ft^3$.

Scales Q, K, S and T, and scales $Q_1$, $K_1$, $S_1$ and $T_1$ are companions and must always be used together. Scale V must always be used with scale S or $S_1$, and scale $V_1$ must always be used with scale T or $T_1$.

### EXAMPLE

What size of electric heater is required to produce a 30°F temperature rise in a system handling 500 cfm of air? How much current will flow in a three-phase circuit at 240 V?

### Solution

On the nomograph, align 500 cfm on scale $Q_1$ with 30°F on scale R, extend the line to scale $K_1$, and read required electric heating coil capacity as 5 kW.

Align 5 kW on scale $K_1$ with 240 V on scale $V_1$, extend the line to scale $T_1$, and read the three-phase current flow as 12 A.

Problems beyond the range of the nomograph can be solved by dividing all known data by a common factor and then multiplying the answer by that factor.

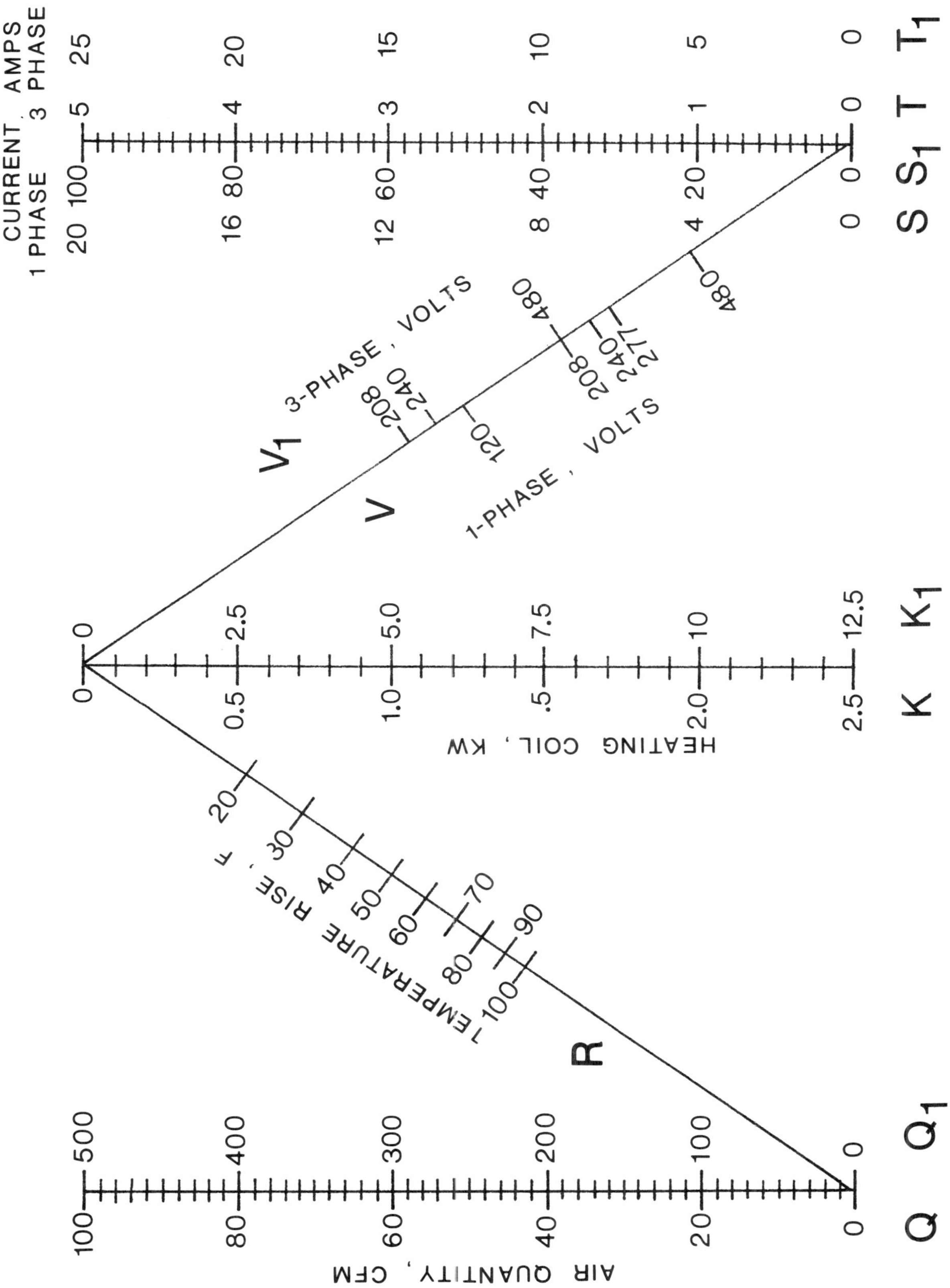
AIR QUANTITY, CFM
Q
Q1
0 20 40 60 80 100
0 100 200 300 400 500
R
TEMPERATURE RISE, F
20 30 40 50 60 70 80 90 100
HEATING COIL, KW
K
K1
0 0.5 1.0 .5 2.0 2.5
0 2.5 5.0 7.5 10 12.5
V
V1
3-PHASE, VOLTS
208 240 480
1-PHASE, VOLTS
120 208 240 277 480
CURRENT, AMPS
1 PHASE
3 PHASE
S
S1
T
T1
0 4 8 12 16 20
0 20 40 60 80 100
0 1 2 3 4 5
0 5 10 15 20 25

## 1-12. Calculating kilowatt requirements

This nomograph makes it possible to quickly and easily solve HVAC electric heating problems. Once the volume of air flow in cfm and the required temperature rise in °F through a heater are known, the required kilowatt rating of the heater can be determined. The kilowatt rating is obtained by the formula:

$$kW = cfm \times T/3193$$

**EXAMPLE**

Find the kW required to raise 2000 cfm through a temperature rise of 10°F.

**Solution**

Align 10 on scale F with 2000 cfm on scale Q and read 6.25 kW on scale KW.

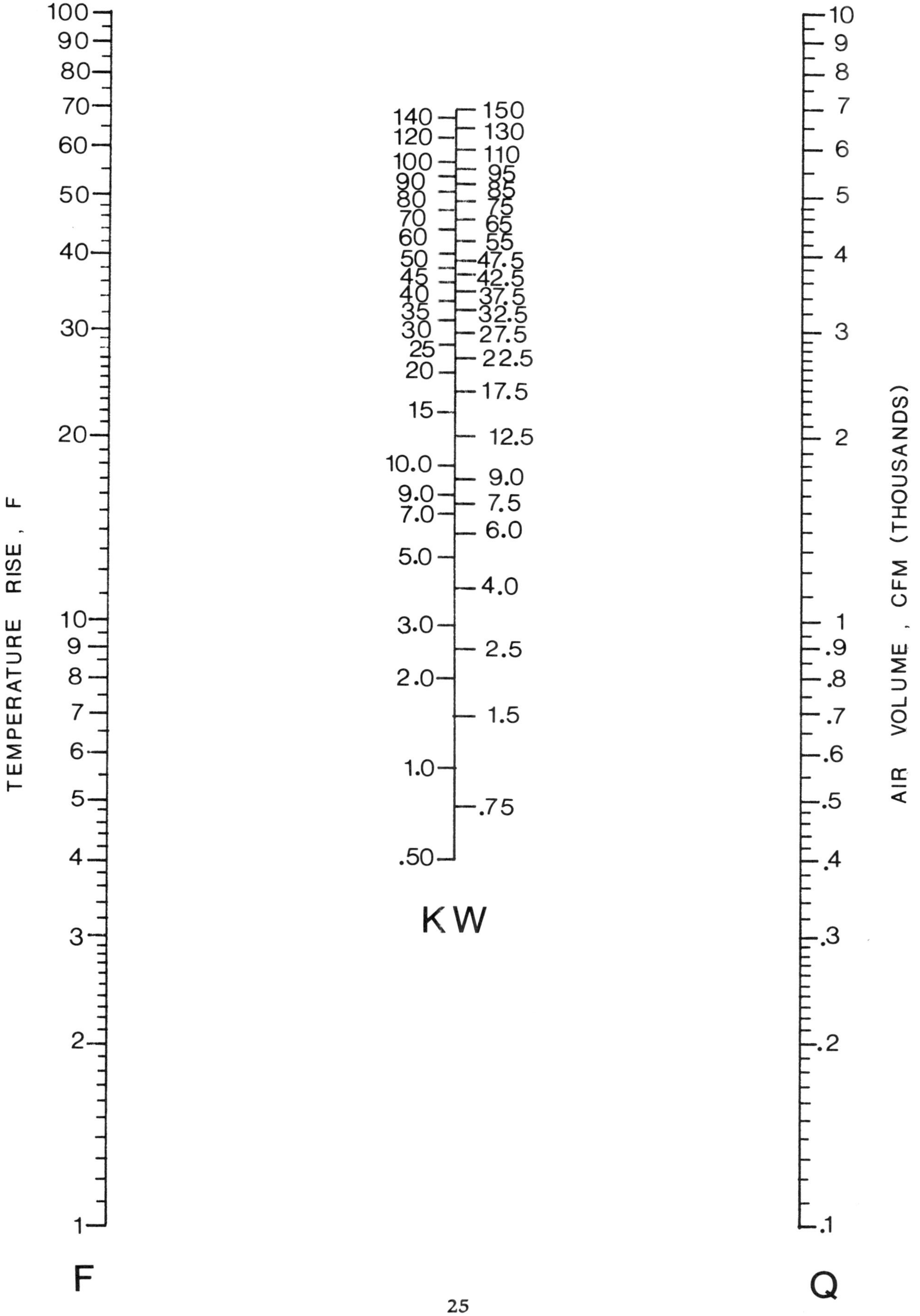
TEMPERATURE RISE , F
100
90
80
70
60
50
40
30
20
10
9
8
7
6
5
4
3
2
1
F
140
120
100
90
80
70
60
50
45
40
35
30
25
20
15
10.0
9.0
7.0
5.0
3.0
2.0
1.0
.50
150
130
110
95
85
75
65
55
47.5
42.5
37.5
32.5
27.5
22.5
17.5
12.5
9.0
7.5
6.0
4.0
2.5
1.5
.75
KW
AIR VOLUME , CFM (THOUSANDS)
10
9
8
7
6
5
4
3
2
1
.9
.8
.7
.6
.5
.4
.3
.2
.1
Q

## 1–13. Calculates heat transfer, water to air

This nomograph makes it possible to quickly and easily solve HVAC water-to-air heat transfer problems in air handling units, unit heaters, fan coil units, etc.

Heat transfer is obtained from the formulas:

$$H = 500 \times Q \times T$$

$$Sh = 1.087 \times V \times T$$

where
- H = heat, Btuh
- Q = water quantity, gpm
- Sh = sensible heat, Btuh
- V = air volume, cfm
- T = temperature change, T

In typical water-to-air heat transfer problems, several parameters are necessary to solve the problem. Depending on the known parameters, the unknown variables can be determined.

### EXAMPLE

Determine the temperature change of air through a 2500-cfm unit heater with a 100 MBtuh capacity and the quantity of water needed when the temperature change of the water is 10°F.

### Solution

Align 2500 cfm on the V scale with 100 MBtuh on the H scale and read 37°F on the TA scale. Next, align 100 MBtuh on the H scale with 10°F on the TW scale and read 20 gpm on the Q scale.

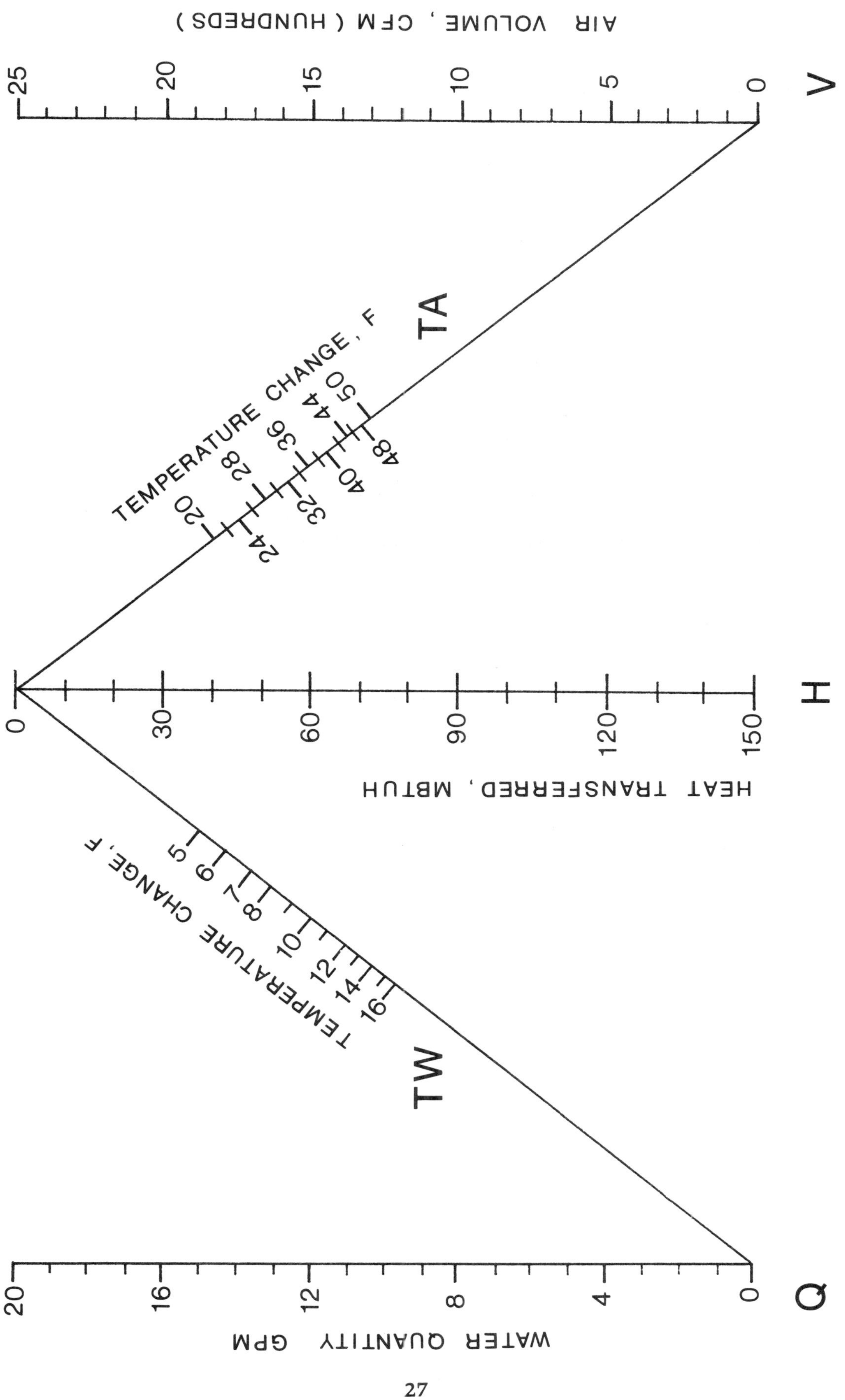
AIR VOLUME, CFM (HUNDREDS)
V
0
5
10
15
20
25
TA
TEMPERATURE CHANGE, F
20
24
28
32
36
40
44
48
50
H
HEAT TRANSFERRED, MBTUH
0
30
60
90
120
150
TW
TEMPERATURE CHANGE, F
5
6
7
8
10
12
14
16
Q
WATER QUANTITY GPM
0
4
8
12
16
20

## 1-14. Calculates heat transfer, steam to air

This nomograph makes it possible to quickly and easily solve HVAC steam to air heat transfer problems for heating air-in-air handling units, unit heaters, etc.

The heat transfer is obtained from the formula:

$$P = \frac{1.087 \times Q \times T}{L}$$

where
- P = steam, lb/hr
- Q = air volume, cfm
- T = temperature change, °F
- L = latent heat of steam, Btu/lb

In typical steam-to-air heat transfer problems, several parameters are necessary to solve the problem. Depending on the known parameters, the unknown variables can be determined.

### EXAMPLE

Determine the volume of air required through a unit heater sized for 100 lb/hr, 25 psig steam when the air temperature change is to be 50°F.

#### Solution

Align 100 lb/hr on scale S with 25 psig on P scale and mark the intersection on H scale. Align the mark on H scale with 50 T on T scale and read 1715 cfm on V scale.

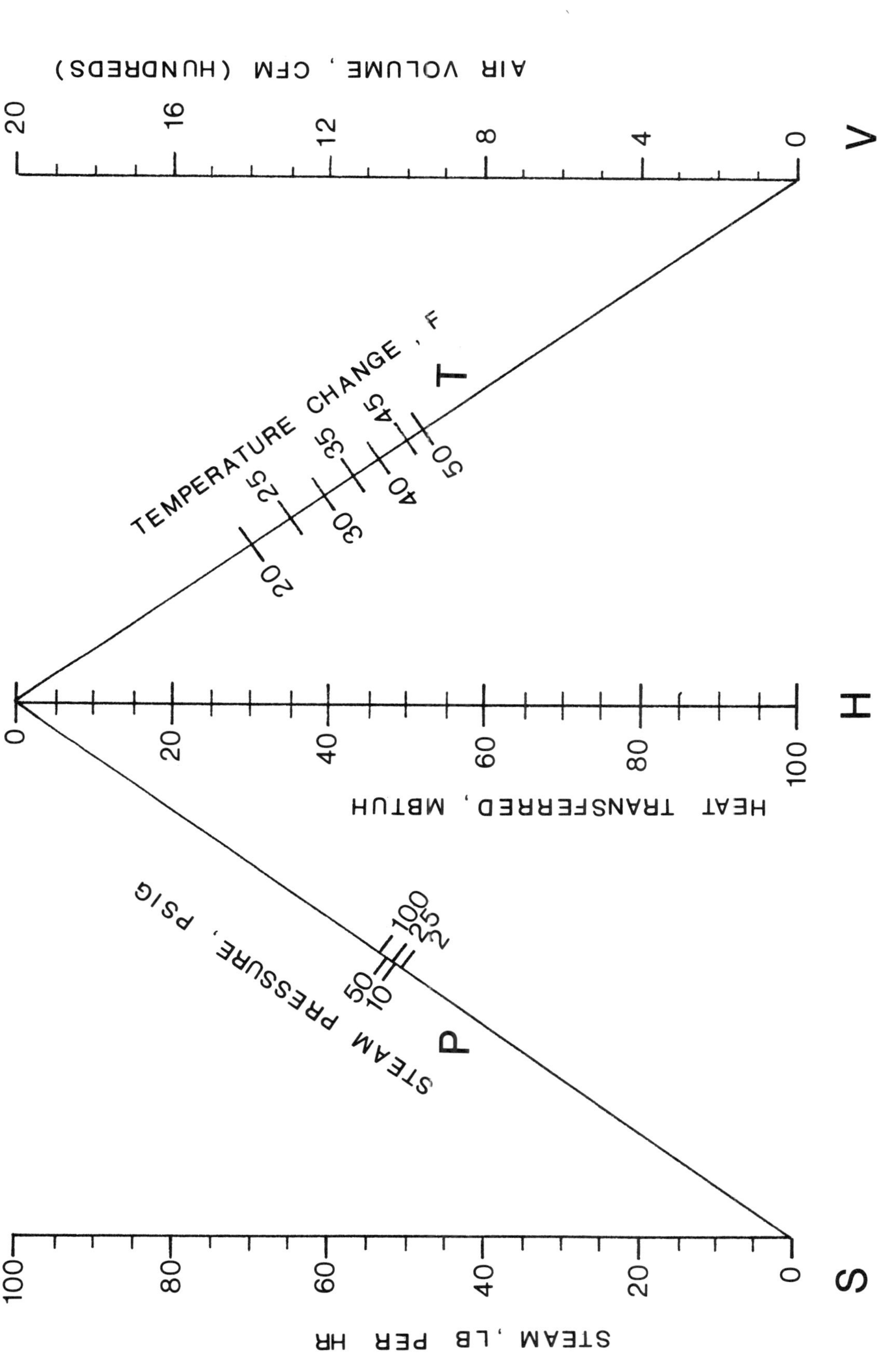
STEAM, LB PER HR
S
0
20
40
60
80
100
STEAM PRESSURE, PSIG
P
10
50
25
100
HEAT TRANSFERRED, MBTUH
H
0
20
40
60
80
100
TEMPERATURE CHANGE, F
T
20
25
30
35
40
45
50
AIR VOLUME, CFM (HUNDREDS)
V
0
4
8
12
16
20

## 1-15. Fan air and brake horsepower calculator

This nomograph solves air handling problems involving fan air and brake horsepower. With it you can quickly and easily determine the size of electric motor required to move a certain quantity of air through a system.

The brake horsepower is obtained from the formula:

$$BHP = Q \times SP/6350 \times ME$$

where
BHP = brake horsepower
SP = static pressure, in. water gauge (WG)
ME = mechanical efficiency
Q = air volume, cfm

The constant 6350 could range from 6346 to 6358, therefore 6350 is a fair compromise and the accuracy is 1 in 600 which is far better than actual field measurement capability.

### EXAMPLE

Determine the size of the motor required to move 5000 cfm of standard air through a system with 5 in. WG static pressure and a mechanical efficiency of 65%.

### Solution

Align 5000 cfm on Q scale with 5 in. WG on P scale and mark the intersection on AH scale. Align this point with 65 on E scale and read 6 hp on BH scale.

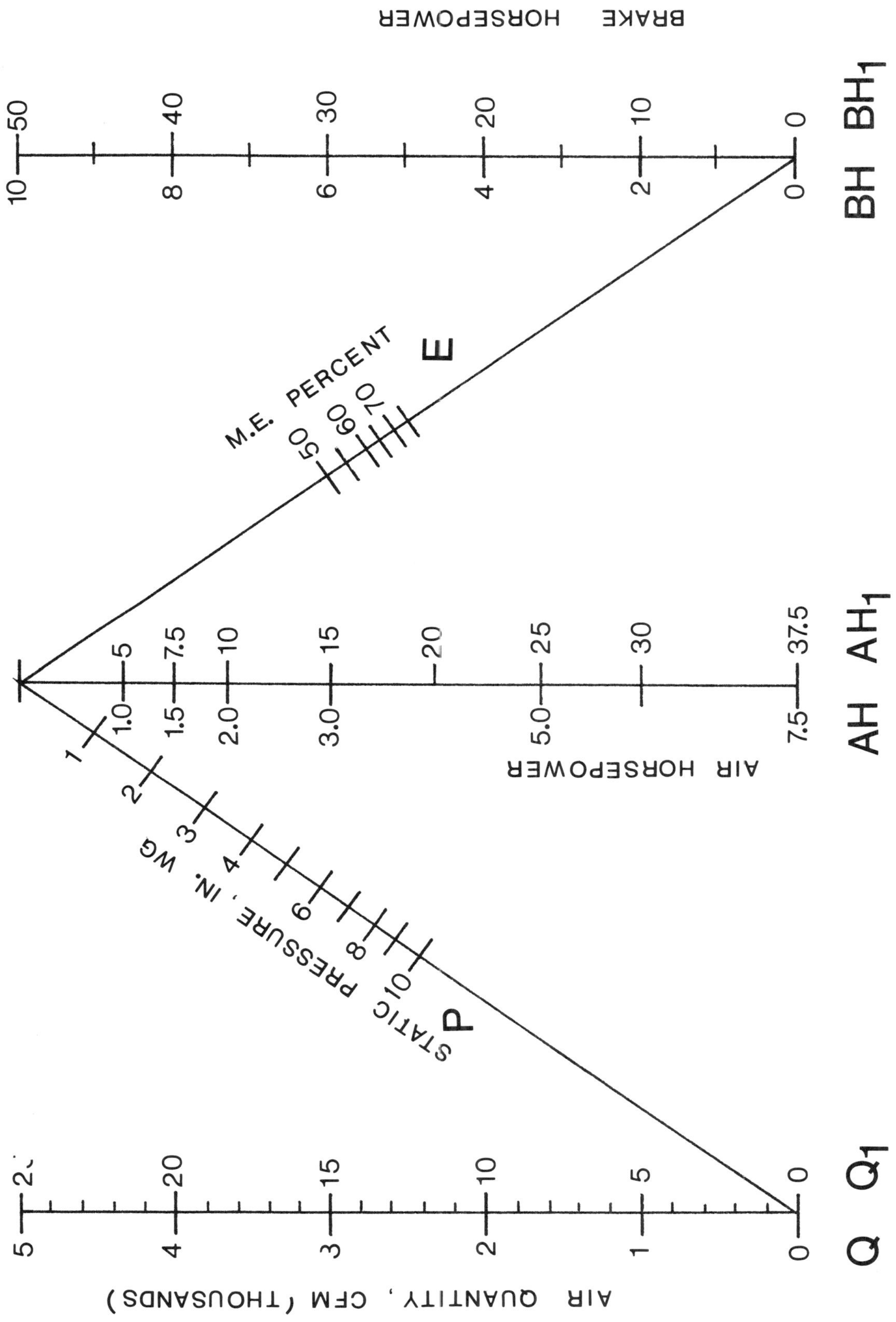
AIR QUANTITY, CFM (THOUSANDS)
Q
Q1
STATIC PRESSURE, IN. WG
P
AIR HORSEPOWER
AH
AH1
M.E. PERCENT
E
BRAKE HORSEPOWER
BH
BH1

# GROUP 2
# WATER

## 2-1. Gives electric water heater input

The nomograph on the following page is useful in solving hot water heater problems when electrical energy is used to heat the water. The electrical energy input rate required to heat water is obtained from the formula:

$$E = 8.33\ QT/3413$$

where E = electrical energy input rate, kW
Q = volume of water heated, gph
T = temperature rise (leaving minus entering), °F

In a typical electric water heater problem, these three parameters are needed to select and schedule the unit properly. With the nomograph on the following page, one can easily determine the unknown parameter.

### EXAMPLE

Determine the size of electric heater, in kW, required to heat water from 50 to 120°F at a rate of 600 gph.

### Solution

Extend a line from 600 on the Q′ scale through 70 on the T scale to the E′ scale, and there read a value of 102.5 kW.

Note that Q and E, and Q′ and E′ are companion scales. Further, the nomograph can be used with parameters beyond the ranges of the scales. For example, if 1200 gph (2 × 600) were used in the above example, the answer would be 205 kW (2 × 102.5). The same is true of the temperature rise; if a 140°F (2 × 70) rise were required, the answer would again be 205 kW (2 × 102.5).

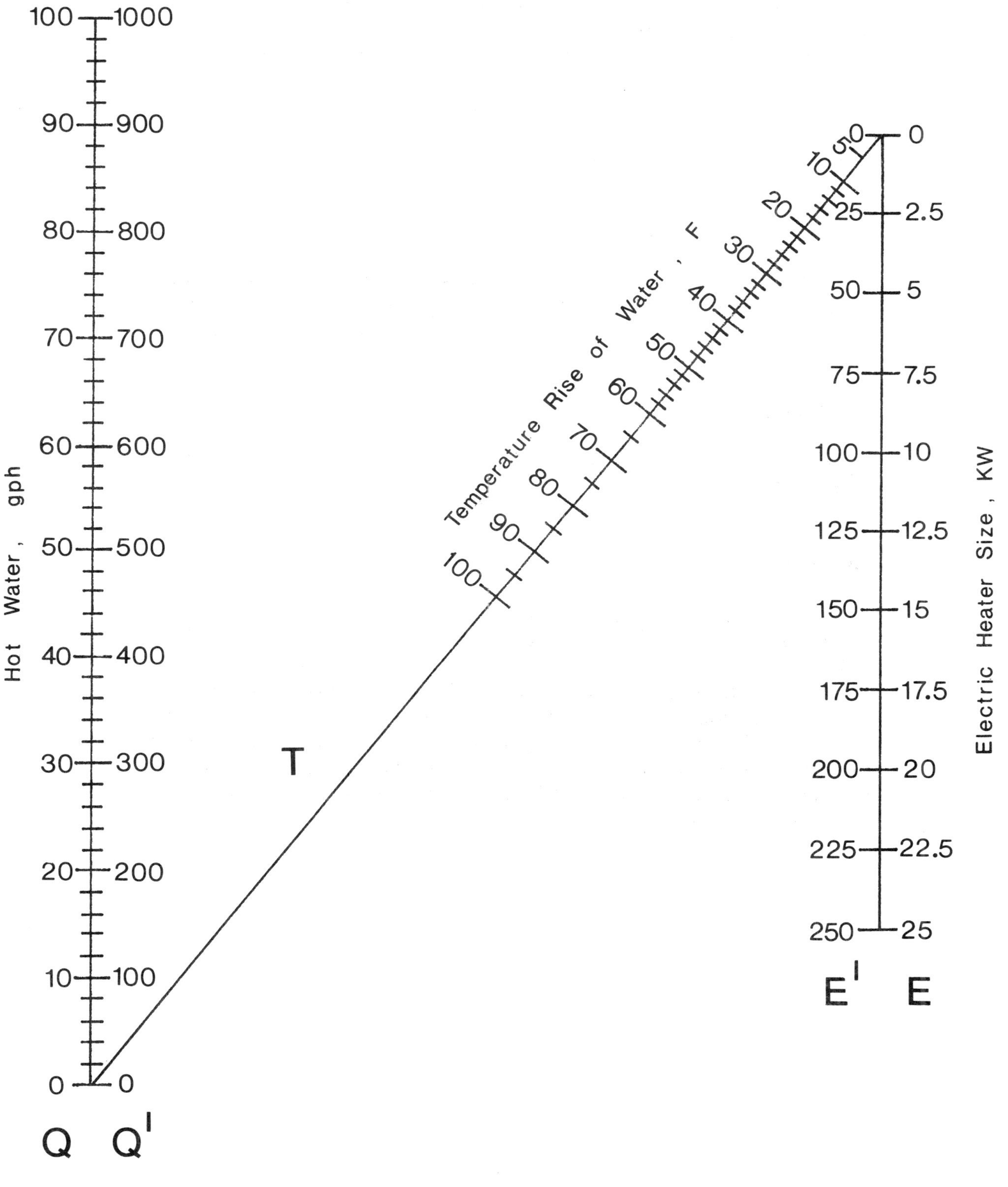
Hot Water, gph
Q
Q'
0
10
20
30
40
50
60
70
80
90
100
0
100
200
300
400
500
600
700
800
900
1000
T
Temperature Rise of Water, F
0
5
10
20
30
40
50
60
70
80
90
100
E'
E
0
25
50
75
100
125
150
175
200
225
250
0
2.5
5
7.5
10
12.5
15
17.5
20
22.5
25
Electric Heater Size, KW

## 2-2. Calculates steam required to heat water in hot water convertor

This nomograph is useful in solving hot water convertor problems where steam is used to heat water.

Steam required to heat water is obtained from the formula:

$$P = gph \times 8.34 \times T/L$$

where
- P = lb/hr steam
- gph = gal/hr water
- T = temperature rise of water T
- L = latent heat of steam

In a typical water heating problem, these four parameters are needed to properly specify and schedule the heating unit. With this nomograph, one can easily and quickly determine the unknown parameter.

### EXAMPLE

Determine the amount of 25 psig steam required to heat water from 50 to 120°F at 800 gph.

### Solution

Align 800 gph on the Q′ scale with 70°F on the temperature rise scale for 25 psig steam, and read the answer on the S′ scale as 500 lb/hr of steam.

Note the scales Q and S, and scales Q′ and S′ are companion scales. Also, this nomograph can be used with parameters beyond the range of the scales. For example, if 1600 gph were used (2 × 800) in the above example, the answer would be 1000 lb/hr (2 × 500). The same is true for the temperature. Suppose 140°F were necessary (2 × 70); the answer would be1000 lb/hr (2 × 500).

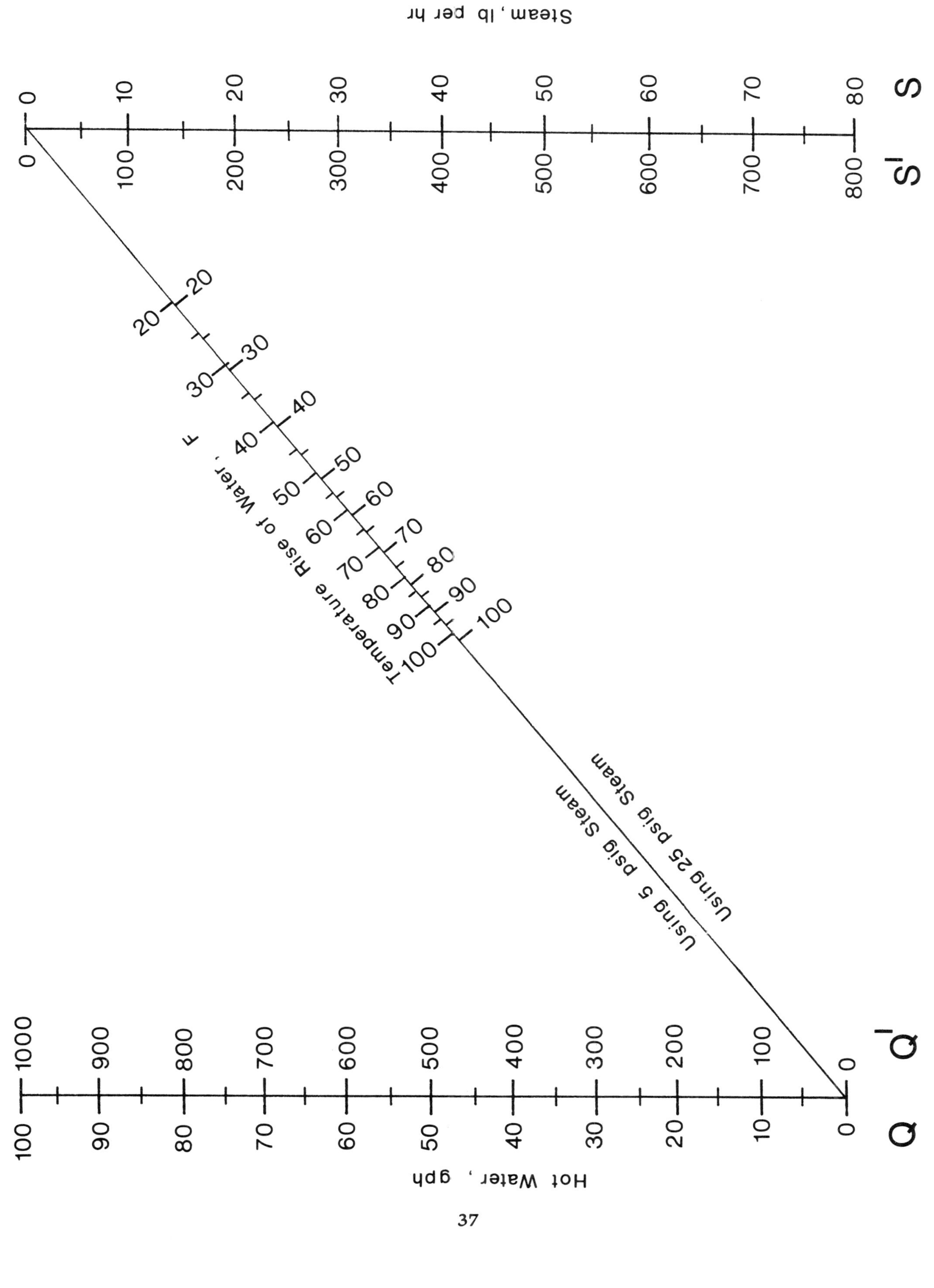
Steam, lb per hr
S
S'
Temperature Rise of Water, F
Using 25 psig Steam
Using 5 psig Steam
Hot Water, gph
Q
Q'

## 2–3. Determines parameters for shell and tube heat exchangers

This simple nomograph makes it possible to quickly and easily solve water heat transfer problems in shell and tube heat exchangers, such as water-cooled condensers and water chillers used in air conditioning systems, or hot water generators (converters) used for domestic water heating and water heating for heating systems.

Heat transfer is obtained from the formula:

$$H = 500 \times Q \times T$$

where
- H = heat, Btuh
- Q = water quantity, gpm
- T = temperature change, °F

In a typical heat transfer problem involving water, three parameters are necessary to solve the problem—the amount of heat transferred, the quantity of water circulated and the temperature change of the water passing through the heat exchanger. With this nomograph and a straightedge, you can quickly determine the unknown parameter when any two are known.

### EXAMPLE

Determine the amount of heat transferred by a shell and tube water chiller with 50 gpm flowing through the unit and the water temperature difference of 12°F.

### Solution

Align 50 gpm on the Q scale with 12°F on the T scale and read 300 on the H scale. Note that the units for the H scale are in MBtuh, so the answer is 300,000 Btuh.

Note that for any problem with parameters beyond the range of this nomograph, the answer can be multiplied by the same factor used on the original parameter. For example, for 500 gpm (which is 50 gpm × 10) at a temperature difference of 12°F, the answer would be 3000 MBtuh (which is 300 MBtuh × 10).

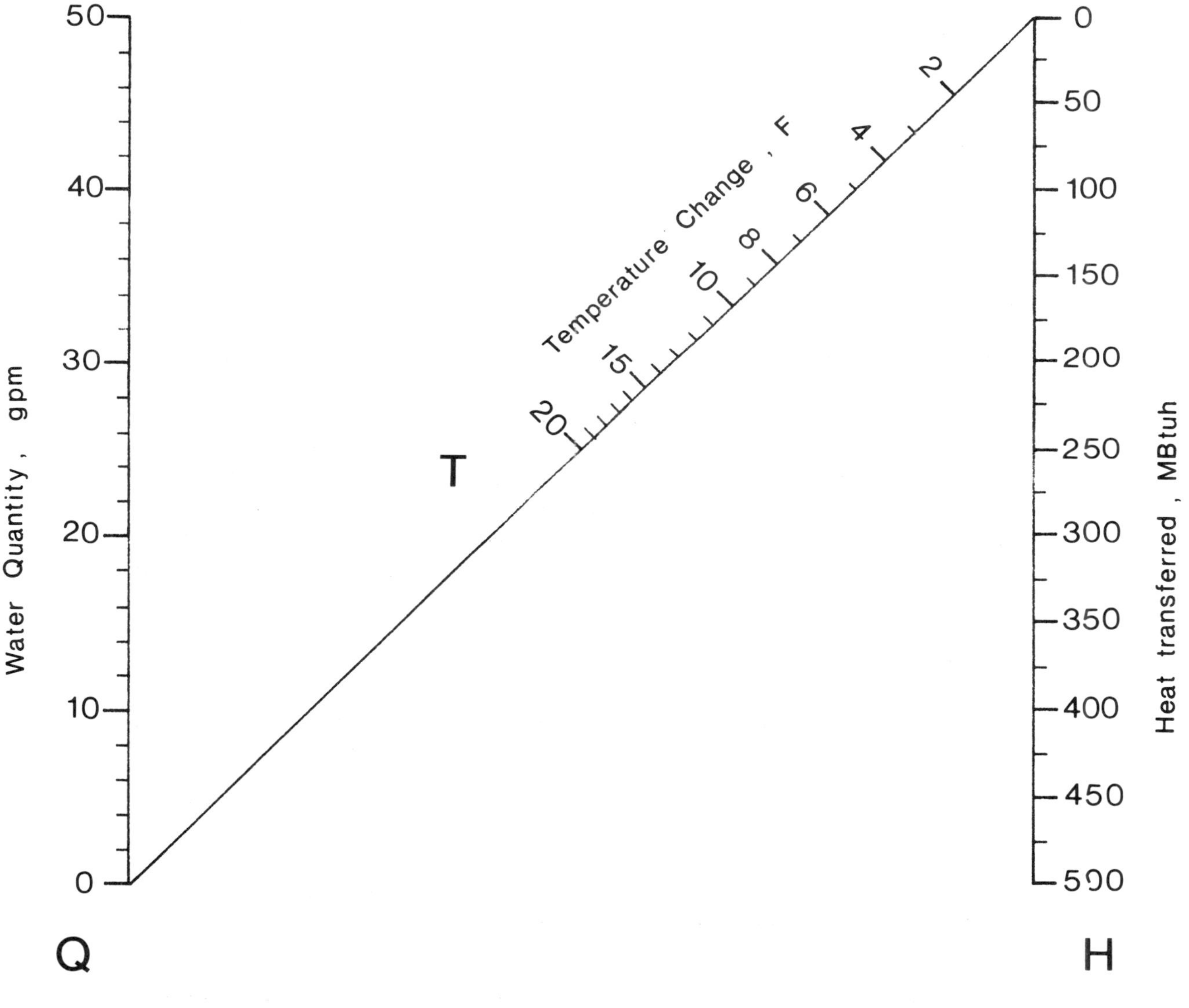
Water Quantity , gpm
50
40
30
20
10
0
Q
T
Temperature Change , F
2
4
6
8
10
15
20
Heat transferred , MBtuh
0
50
100
150
200
250
300
350
400
450
500
H

## 2-4. Determines mean velocity of flowing liquids in pipes

This nomograph is useful in determing the rate of liquid flow, mean velocity of flow or nominal pipe diameter.

The mean velocity of any flowing liquid is obtained from the formula:

$$v = 0.408 \times Q/d^2$$

where $v$ = mean velocity, fps
$Q$ = rate of flow, gpm
$d$ = internal diameter of pipe, in. (Schedule 40 pipe)

In a typical liquid flow problem where any two of the three parameters are known, the other one can easily and quickly be determined with this nomograph.

### EXAMPLE

Determine the mean velocity of water flowing at a rate of 45 gpm through a 2-in. Schedule 40 pipe.

#### Solution

Align 45 gpm on the Q scale with 2 in. on the D scale and read the answer as 4.3 fps on the V scale.

Note that scales Q, D and V are companion scales with Q′, D′ and V′. In cases where the velocity must be maintained below some maximum value, align that number with the known flowrate and select the next larger pipe size. For example, what pipe size is needed when the maximum velocity must be below 10 fps with an 800-gpm flowrate? To solve, align 800 gpm on the Q′ scale with 10 fps on the V′ scale and select the next larger pipe on the D′ scale, 6 in.

### OTHER USEFUL INFORMATION

The following is intended to serve as a guide for recommended good practice to keep fluid velocities within a reasonable range:

- commercial buildings—occupied space (quiet zones), 4 to 6 fps maximum velocity; mechanical rooms, penthouses, etc., 8 fps maximum velocity; and
- industrial buildings—10 to 12 fps maximum velocity.

Shell and tube heat exchanger tube velocity above a maximum of 7.5 fps can be erosive. The following are the rates of flow ranges by pipe size to keep velocities within a reasonable range.

| Nominal Pipe Size (in.) | Flow Range (gpm) | Nominal Pipe Size (in.) | Flow Range (gpm) |
|---|---|---|---|
| 1.5 | 10-25 | 4 | 125-175 |
| 2 | 20-50 | 5 | 200-500 |
| 2.5 | 30-75 | 6 | 325-800 |
| 3 | 50-140 | 8 | 700-1300 |
| | | 10 | 1250-1750 |

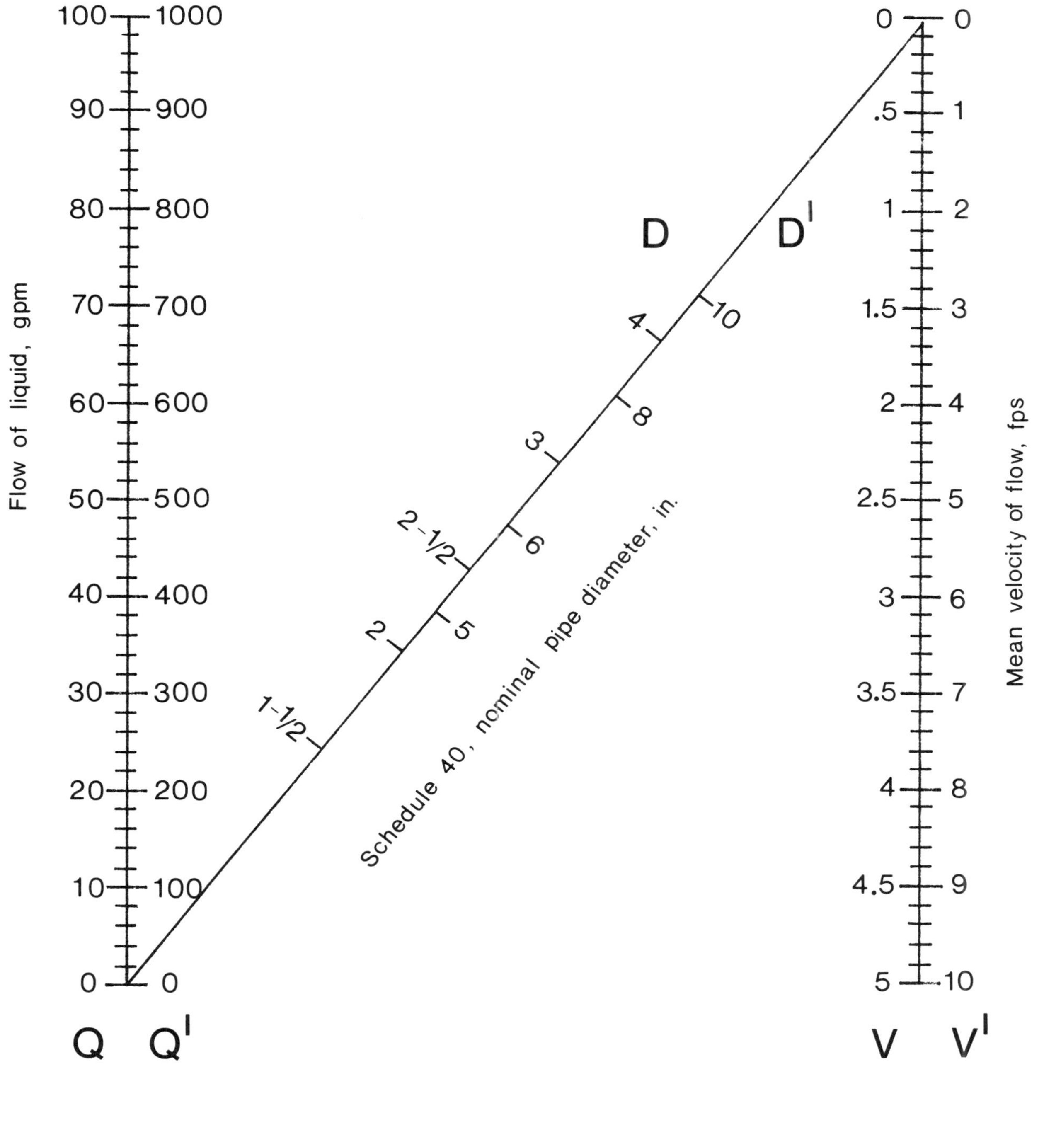

Flow of liquid, gpm
100
1000
90
900
80
800
70
700
60
600
50
500
40
400
30
300
20
200
10
100
0
0
Q
Q'
D
D'
1-1/2
2
2-1/2
3
4
5
6
8
10
Schedule 40, nominal pipe diameter, in.
0
0
.5
1
1
2
1.5
3
2
4
2.5
5
3
6
3.5
7
4
8
4.5
9
5
10
V
V'
Mean velocity of flow, fps

## 2–5. Determines quantity of water in steel pipe

In many water piping systems it is necessary to know the quantity of water in the system for filling and draining purposes. This nomograph is useful in determining the quantity of water in gallons for pipe sizes ranging from 4 to 24 in. diameter and in any length. The nomograph is designed for Schedule 40 steel pipe using nominal pipe diameter in inches.

### EXAMPLE

Determine the quantity of water in gallons contained in a 20-in. condenser-water vertical pipe stack 25 ft high.

### Solution

Align 25 ft on scale L with 20 in. on scale D′ and read 360 gal on scale Q′. Note: Scales D and Q, and D′ and Q′ are companion scales, i.e., for pipe sizes 4–12 in. on scale D read answer on scale Q and for pipe sizes 8–24 in. on scale D′ read answer on scale Q′.

For any pipe length beyond the range of scale L, divide the actual length by a number that yields a value within the range of scale L. Determine the quantity of water as above, then multiply that answer by the same factor used in the division and the result will be the actual quantity of water in the pipe.

For example, if the length in the problem were 250 ft, dividing 250 by 10 yields 25, solve as before, then multiply 360 by 10 and the result is 3600 gal.

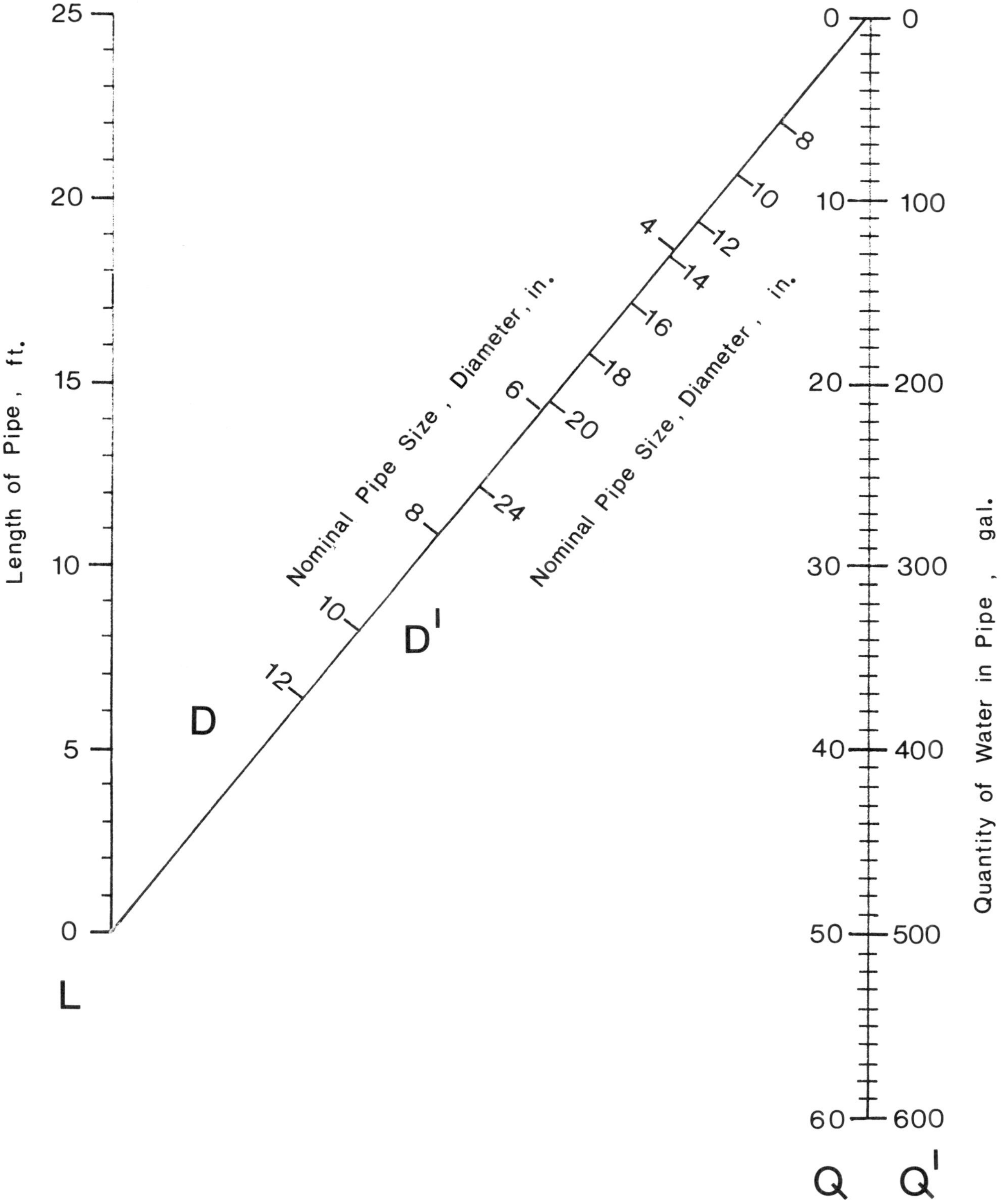

Length of Pipe , ft.
25
20
15
10
5
0
L
D
D'
Nominal Pipe Size , Diameter , in.
12
10
8
6
4
Nominal Pipe Size , Diameter , in.
8
10
12
14
16
18
20
24
0
10
20
30
40
50
60
Q
0
100
200
300
400
500
600
Q'
Quantity of Water in Pipe , gal.

## 2-6. Determines weight of water-filled steel pipe

When designing piping systems, it is necessary to determine the weight of the water-filled pipe so that the correct and economical selection of hangers and supports can be made. This nomograph is designed for Schedule 40 steel pipe using nominal pipe diameter in inches.

### EXAMPLE

Determine the weight of a filled 20-in. Schedule 40 condenser-water vertical steel pipe 25 ft from the basement chiller to the cooling tower on the roof.

### Solution

Align 25 ft on scale L with 20 in. on scale D′ and read 6000 lb on scale W′. Note: Scales D and W, and D′ and W′ are companion scales, i.e., for pipe sizes 4–12 in. on scale D read answer on scale W and for pipe sizes 10–24 in. on scale D′ read answer on scale W′.

For any pipe length beyond the range of scale L, divide the actual length by a number that yields a value within the range of scale L. Determine the weight as above, then multiply that answer by the same factor used in the division and the result will be the actual weight of the water-filled pipe.

For example, if the length in the problem were 100 ft, dividing 100 by 4 yields 25, solve as before, then multiply 6000 by 4 and the result is 24,000 lb.

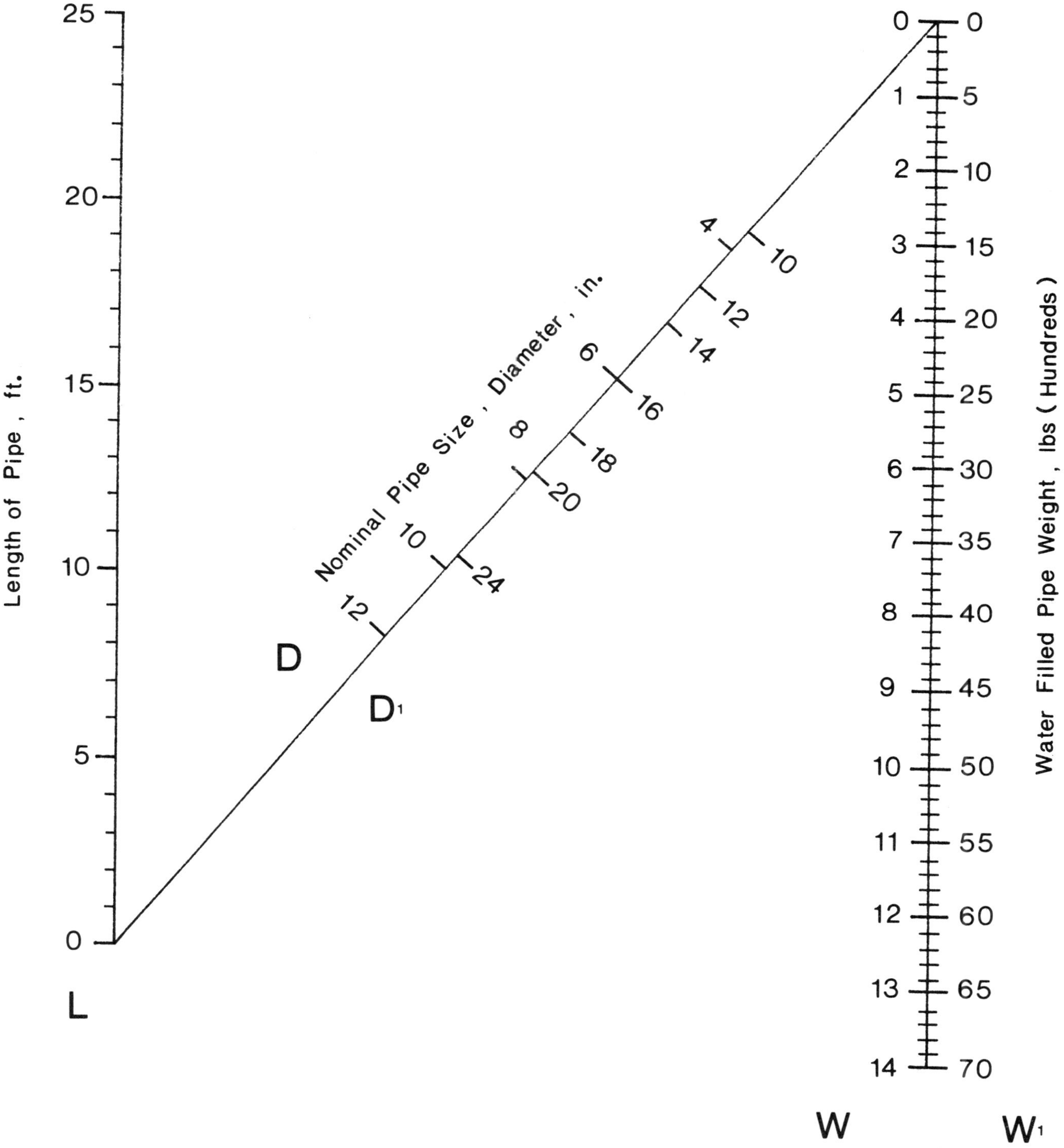
Length of Pipe , ft.
25
20
15
10
5
0
L
Nominal Pipe Size , Diameter , in.
D
D1
4
6
8
10
12
10
12
14
16
18
20
24
Water Filled Pipe Weight , lbs ( Hundreds )
W
W1
0
1
2
3
4
5
6
7
8
9
10
11
12
13
14
0
5
10
15
20
25
30
35
40
45
50
55
60
65
70

## 2–7. Determines quantity of water in copper pipes

This nomograph is useful in determining the quantity of water in gallons for pipe sizes ranging from 1 to 6 in. diameter in any length. The nomograph is designed for Type "K" hard copper pipe using nominal diameter in inches.

### EXAMPLE

Determine the quantity of water in gallons contained in a 4-in. pipe 50 ft long.

### Solution

Align 50 ft on scale L with 4 in, on scale D′ and read 30 gal on scale Q′. Note: Scales D and Q, and D′ and Q′ are companion scales, i.e., for pipe sizes 1–3 in. on scale D read answer on scale Q and for pipe sizes 3–6 in. on scale D′ read answer on scale Q′.

For any pipe length beyond the range of scale L, divide the actual length by a number that yields a value within the range of scale L. Determine the quantity of water as above, then multiply that answer by the same factor used in the division and the result will be the actual quantity of water in the pipe.

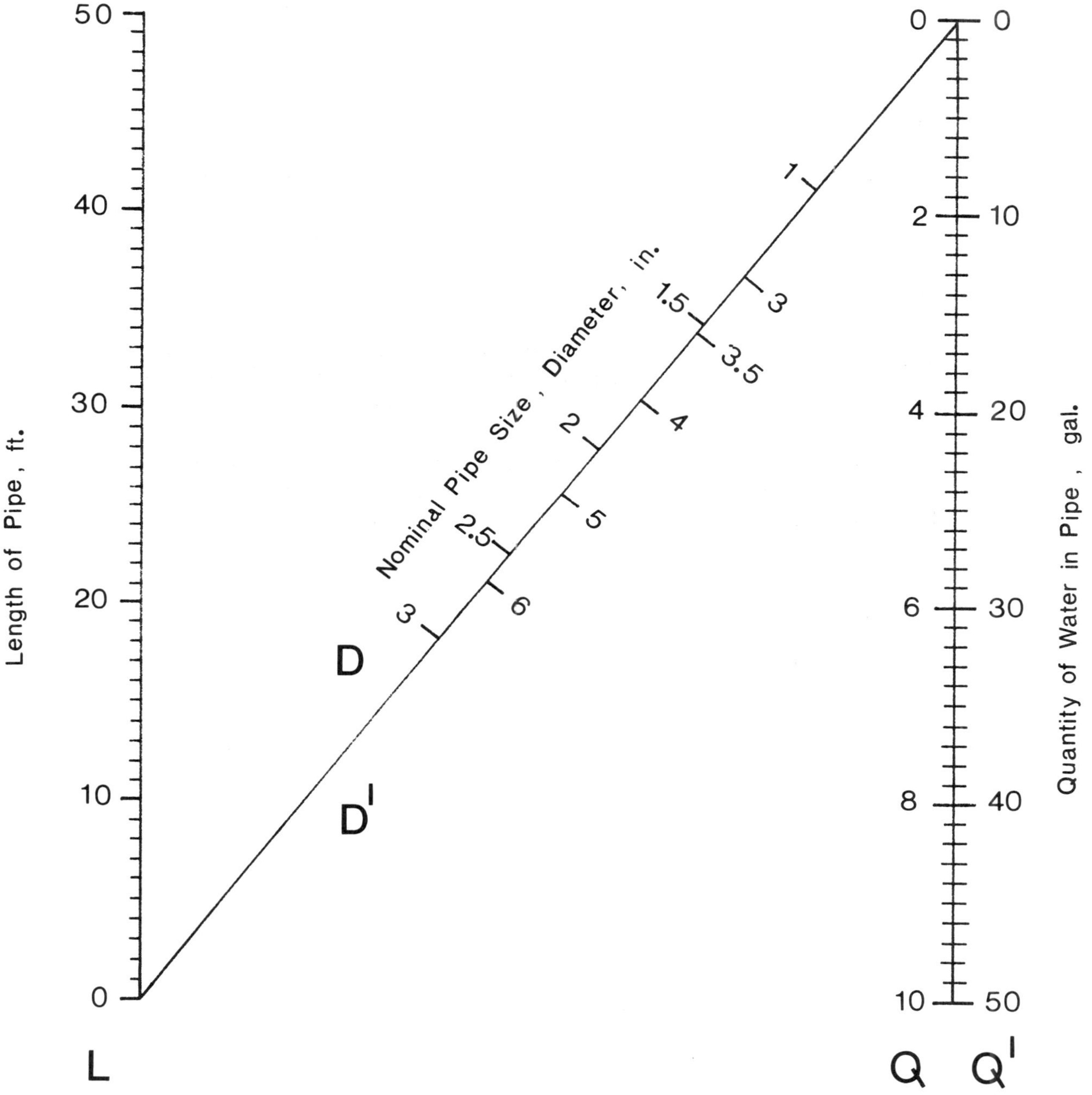
Length of Pipe, ft.
50
40
30
20
10
0
L
Nominal Pipe Size, Diameter, in.
D
D'
1
1.5
2
2.5
3
3
3.5
4
5
6
Quantity of Water in Pipe, gal.
0
2
4
6
8
10
Q
0
10
20
30
40
50
Q'

## 2-8. Determines weight of water-filled copper pipe

When designing piping systems, it is necessary to determine the weight of the water-filled pipe so that the correct and economical selection of hangers and supports can be made. This nomograph is designed for Type "K" hard copper pipe using nominal diameter in inches.

### EXAMPLE

Determine the weight of a vertical 6-in. water-filled copper pipe 50 ft high.

### Solution

Align 50 ft on scale L with 6 in. on scale D′ and read 1000 lb on scale W′. Note: Scales D and W, and D′ and W′ are companion scales, i.e., for pipe sizes 1–3.5 in. on scale D read answer on scale W and for pipe sizes 3–6 in. on scale D′ read answer on scale W′.

For any pipe length beyond the range of scale L, divide the actual length by a number that yields a value within the range of scale L. Determine the weight as above, then multiply that answer by the same factor used in the division and the result will be the actual weight of the water-filled pipe.

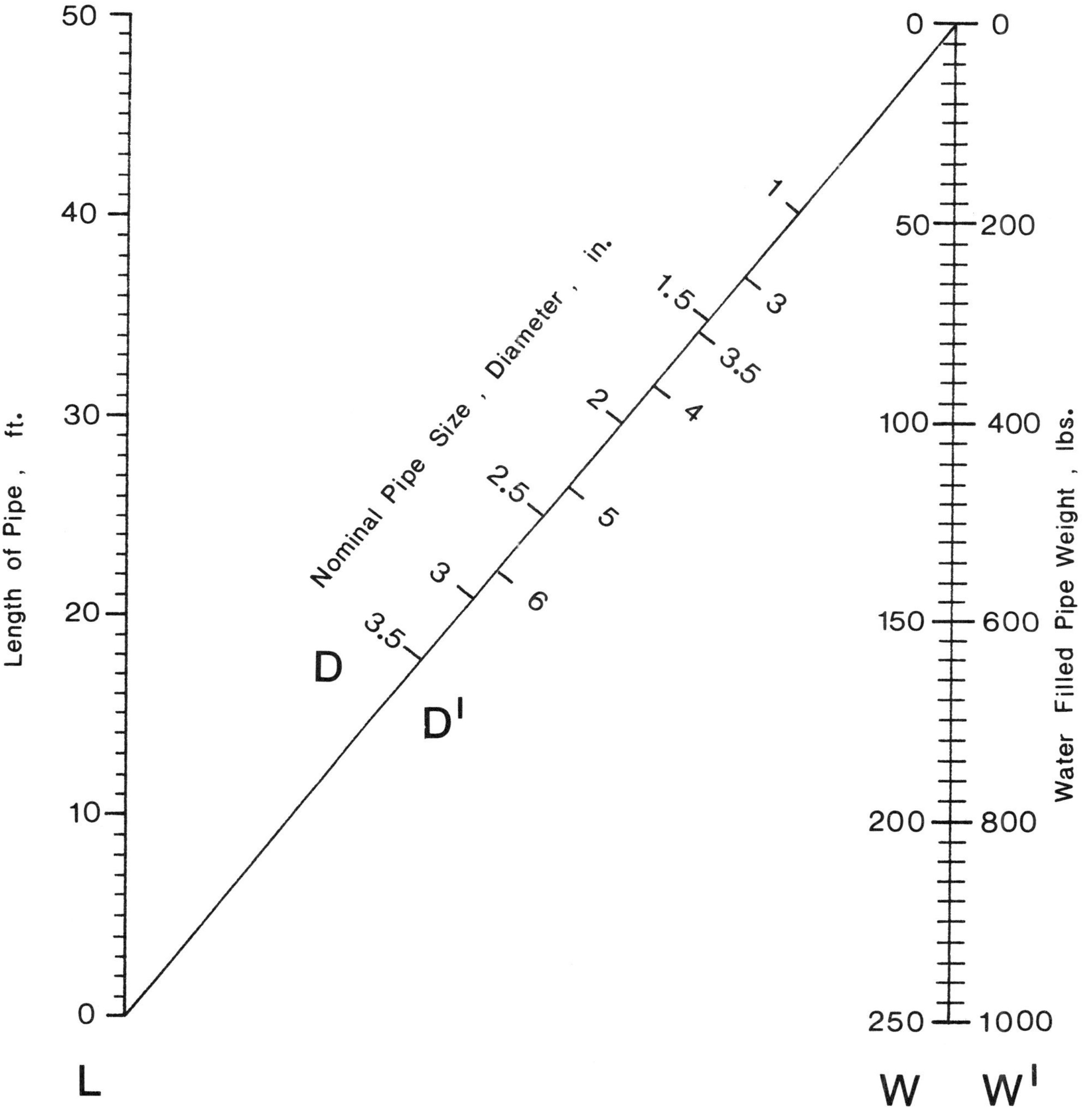
50
40
30
20
10
0
Length of Pipe , ft.
L
Nominal Pipe Size , Diameter , in.
D
D'
1
1.5
2
2.5
3
3.5
3
3.5
4
5
6
0
50
100
150
200
250
W
0
200
400
600
800
1000
W'
Water Filled Pipe Weight , lbs.

## 2-9. Thermal expansion of steel pipe calculator

The linear expansion and contraction of steel pipe due to differences of temperature, for example, between installation and operating temperatures, can be quickly and easily determined with this nomograph. This change in length must be cared for by suitable expansion joints, expansion loops or changes in direction of the pipe.

This linear expansion is obtained from the formula:

$$E = 12 \times C \times L \times T$$

where
- $E$ = change in length, in.
- $C$ = coefficient of linear expansion (for steel averages about $6.7 \times 10^{-6}$ in./in./°F.)
- $L$ = length of pipe, ft
- $T$ = temperature change, °F

### EXAMPLE

Determine the expansion of steel pipe 100 ft long with a temperature change of 300°F.

### Solution

Align 100 ft on scale L with 300°F on scale T and read 2.4 in. on scale E. Note: Scales L and E, and L′ and E′ are companion scales and must be used together on problems.

The coefficient of linear expansion changes with temperature, but for the range of temperatures of this nomograph, an average value is indicated in the formula while the nomograph adjusts for the change.

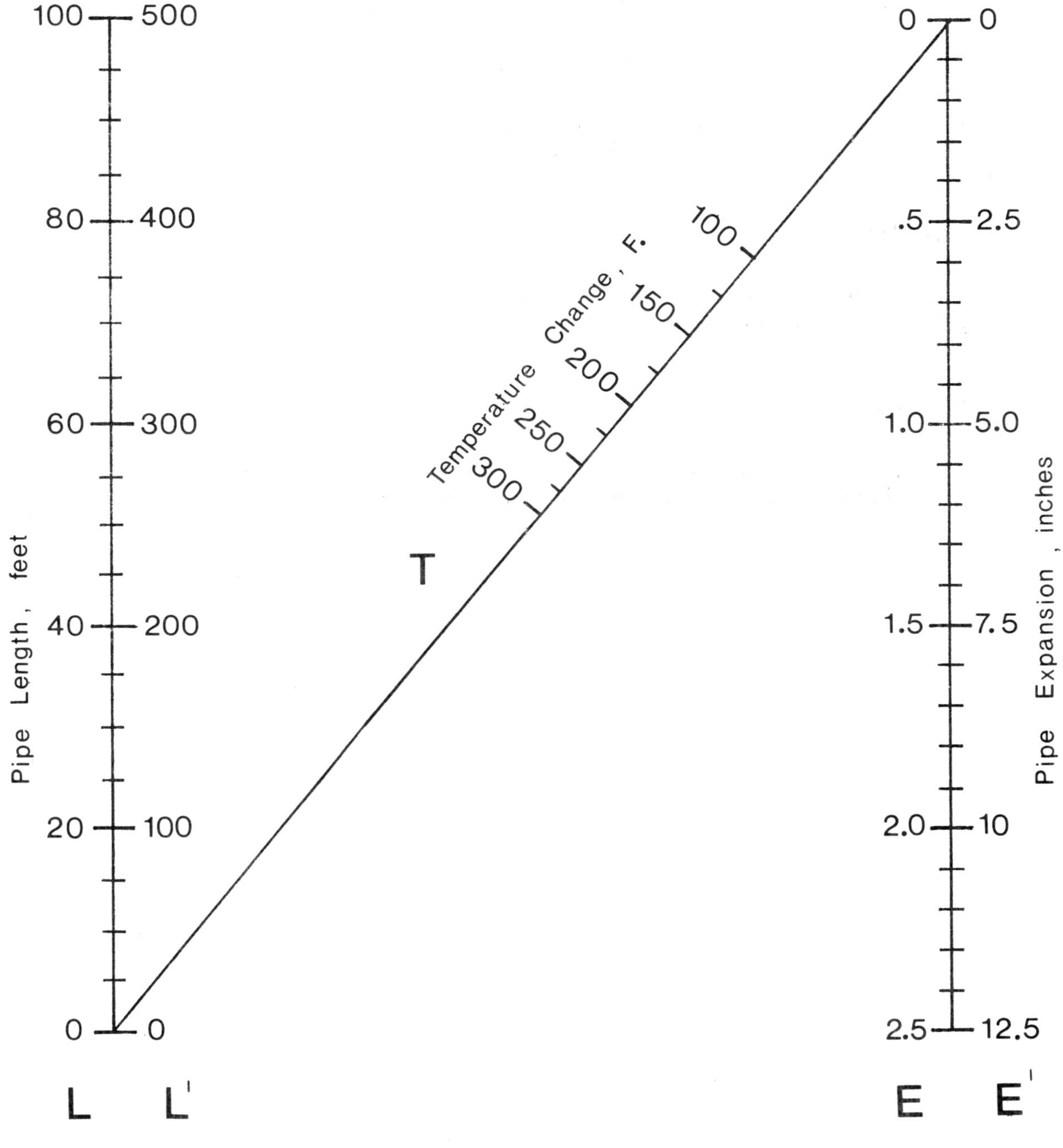
100
80
60
40
20
0
500
400
300
200
100
0
Pipe Length, feet
L
L'
Temperature Change, F.
100
150
200
250
300
T
0
.5
1.0
1.5
2.0
2.5
0
2.5
5.0
7.5
10
12.5
Pipe Expansion, inches
E
E'

## 2-10. Thermal expansion of copper pipe calculator

The linear expansion and contraction of copper pipe due to differences of temperature, for example, between installation and operating temperature, can be quickly and easily determined with this nomograph.

The linear expansion is obtained from the formula:

$$E = 12 \times C \times L \times T$$

where C = coefficient of linear expansion (for copper averages about $9.5 \times 10^{-6}$ in./in./°F)

### EXAMPLE

Determine the expansion of copper pipe 400 ft long with a temperature change of 200°F.

### Solution

Align 400 ft on scale L′ with 200°F on scale T and read 9 in. on scale E′. Note: Formula description and note same as 2-9.

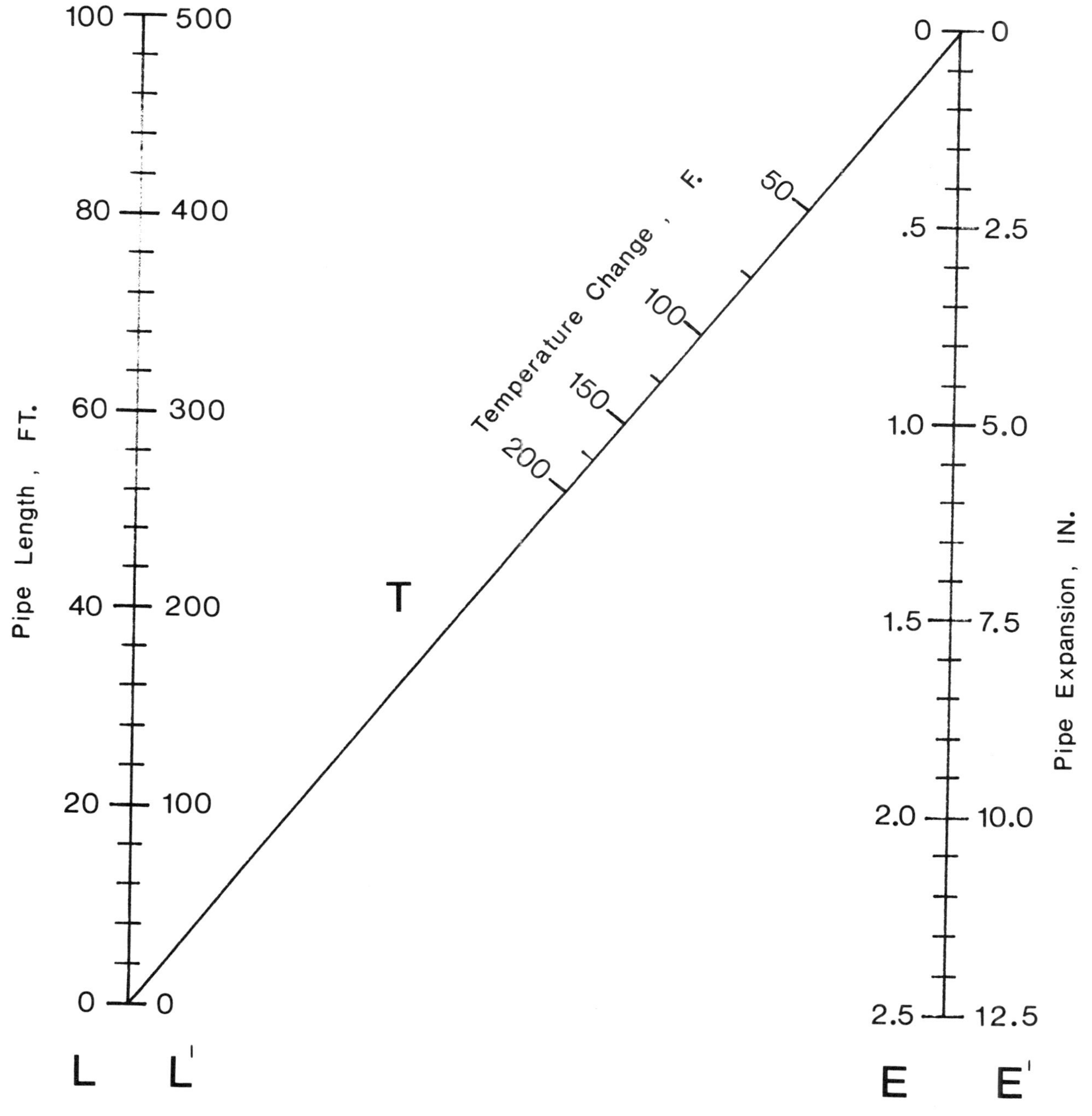
Pipe Length , FT.
100
80
60
40
20
0
500
400
300
200
100
0
L
L'
T
Temperature Change , F.
50
100
150
200
0
.5
1.0
1.5
2.0
2.5
0
2.5
5.0
7.5
10.0
12.5
Pipe Expansion , IN.
E
E'

## 2-11. Solves pump horsepower/water flow problems

This nomograph is useful in determining the brake horsepower or the quantity of water flowing in water flow problems.

Brake horsepower is obtained from the formula:

$$BHP = Q \times H \times S/(3960 \times E)$$

where
- BHP = brake horsepower
- Q = water flow, gpm
- H = total dynamic head, ft of water
- S = specific gravity (S = 1 for water)
- E = pump efficiency

In a typical water flow problem, three parameters are necessary to solve the problem—total dynamic head, water flow and brake horsepower. The unknown parameter can quickly be determined when any two are known.

### EXAMPLE

Determine the brake horsepower required to pump 250 gpm against a 100-ft head with a 70% pump efficiency.

#### Solution

Align 250 gpm on the F′ scale with 100 ft on H scale, and extend this line to reference line (step 1). Align this point on reference line with 0.7 on E scale and read the brake horsepower on HP′ as 9.0 BHP. Note: Scales F and HP, and F′ and HP′ are companion scales and must always be used together. For problems with parameters (gpm, ft and bhp) beyond the range of the scales, the answer must be multiplied by the same factor used on the original parameter. For example, for 500 gpm (250 gpm × 2) at 100 ft head, the answer is 18.0 bhp (9.0 bhp × 2). The same reasoning applies to the dynamic head and brake horsepower scales.

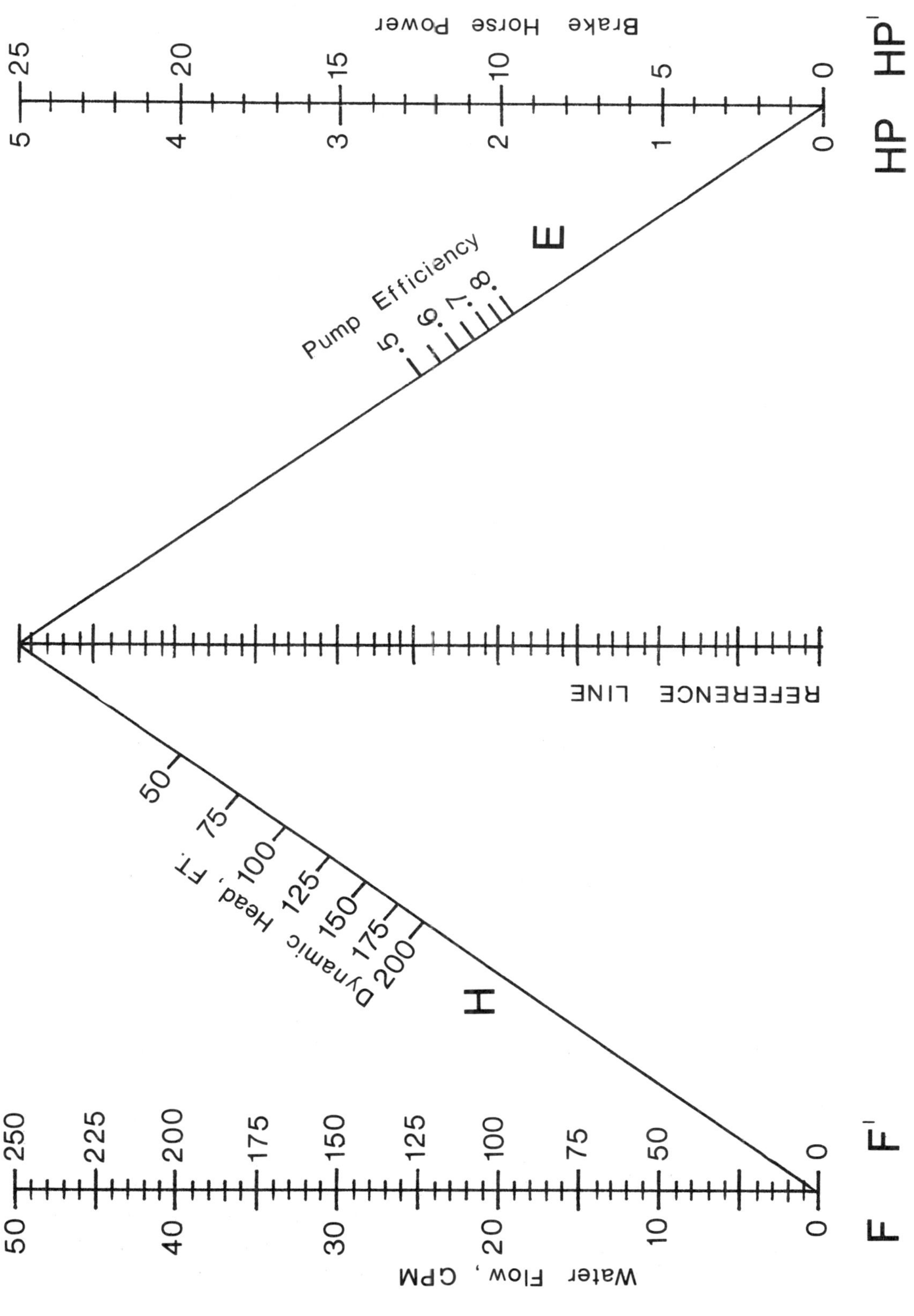
Water Flow, GPM
F
F'
0
10
20
30
40
50
0
50
75
100
125
150
175
200
225
250
H
Dynamic Head, FT.
50
75
100
125
150
175
200
REFERENCE LINE
Pump Efficiency
.5
.6
.7
.8
E
HP
HP'
0
1
2
3
4
5
0
5
10
15
20
25
Brake Horse Power

## 2–12. Quick sizing of building drains

The building drain is the lowest horizontal piping of a drainage system within a building. This drain receives the discharge from soil and waste pipes and conveys it to the building sewer which extends to a public sewer or other disposal areas.

The size of building drain and sewer is determined by the number and types of plumbing fixtures discharging into the system. To have a uniform method for rating plumbing fixtures, the "fixture unit" was established as the unit of measure of the waste produced by the plumbing fixture.

The more common fixtures used in industry are:

| Fixture | Fixture-Unit Value |
|---|---|
| Drinking fountain | 0.5 |
| Lavatory (1.25- or 1.5-in. trap) | 1 |
| Floor drain with 3-in. trap | 1 |
| Service sink (slop) | 3 |
| Urinal-stall or washout | 2 |
| Urinal-pedestal, siphon jet or blow out | 8 |
| Water closet—tank operated | 4 |
| Water closet—flush valve operated | 8 |

Fixture-unit values for fixtures not shown above can be estimated, based on drain or trap size:

| Fixture Drain or Trap Size (in.) | Fixture-Unit Value |
|---|---|
| 1.25 or smaller | 1 |
| 1.5 (except lavatory) | 2 |
| 2 | 3 |
| 2.5 | 4 |
| 3 | 5 |
| 4 | 6 |

To determine fixture-unit value for fixtures or water operated devices operating intermittently, let

$$\text{fixture-unit value} = \text{gpm}/7.5$$

For continuous or semicontinuous flow from a pump, pump ejector, air conditioning equipment or similar devices, allow two fixture-units for each gpm flow.

### EXAMPLE

Determine the size of pipe required for the building drain and sewer when the fixture units are calculated to be 1000 and the fall is ⅛ in./ft.

### Solution

From 1000 on scale C extend a straight line through ⅛ in. on scale F and intersect scale B between 6 and 8 in. pipe. The larger size must be used. However, with a 0.25-in. fall, a 6-in. pipe is satisfactory. Note: Scales A and D and scales B and C are companion scales and must be used together. Values are based on the New Jersey Plumbing Code.

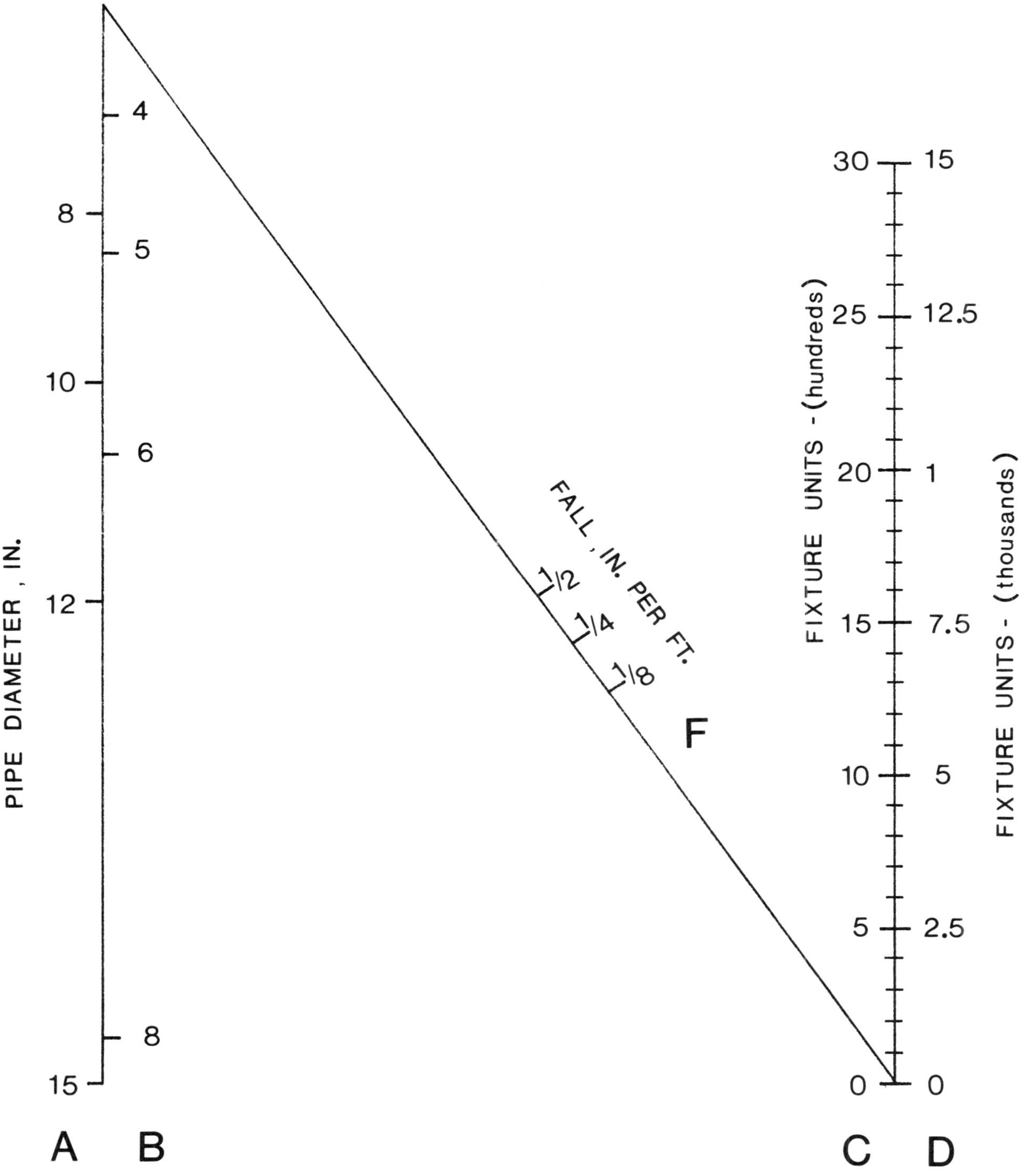
PIPE DIAMETER, IN.
8
10
12
15
4
5
6
8
FALL, IN. PER FT.
1/2
1/4
1/8
F
FIXTURE UNITS - (hundreds)
30
25
20
15
10
5
0
FIXTURE UNITS - (thousands)
15
12.5
1
7.5
5
2.5
0
A
B
C
D

## 2-13. How to determine storm drain pipe size

Optimum pipe size for a storm drain required for draining plant roof areas can be easily and quickly calculated using a nomograph. The nomograph is based on the horizontal projected roof area. A maximum rainfall of 4 in./hr is also assumed; for localities where U.S. Weather Bureau statistics indicate a maximum rainfall other than 4 in./hr, adjust the projected roof area using the formula:

$$ARA = \text{measured roof area} \times \text{actual rainfall}/4$$

Use this roof area figure to find the size of drain pipe required. In cases where vertical walls drain onto the roof, add one-half the wall area to the roof area.

### EXAMPLE

Determine the size of drain pipe required for a roof area of 40,000 $ft^2$, if the drain pipe slope is 0.25 in./ft.

### Solution

Align 40,000 $ft^2$ on scale D with 0.25-in. and read drain pipe size on scale A. In this example, the value lies between 10 and 12 in.; use the larger size, 12.

Note that with a 0.25-in. slope in the drain line, a 10-in. diameter pipe is adequate to carry the flow. Therefore, whenever possible, use the maximum slope and design for optimum, rather than maximum, pipe size.

Scales A and D, and B and C are companion scales and must be used together.

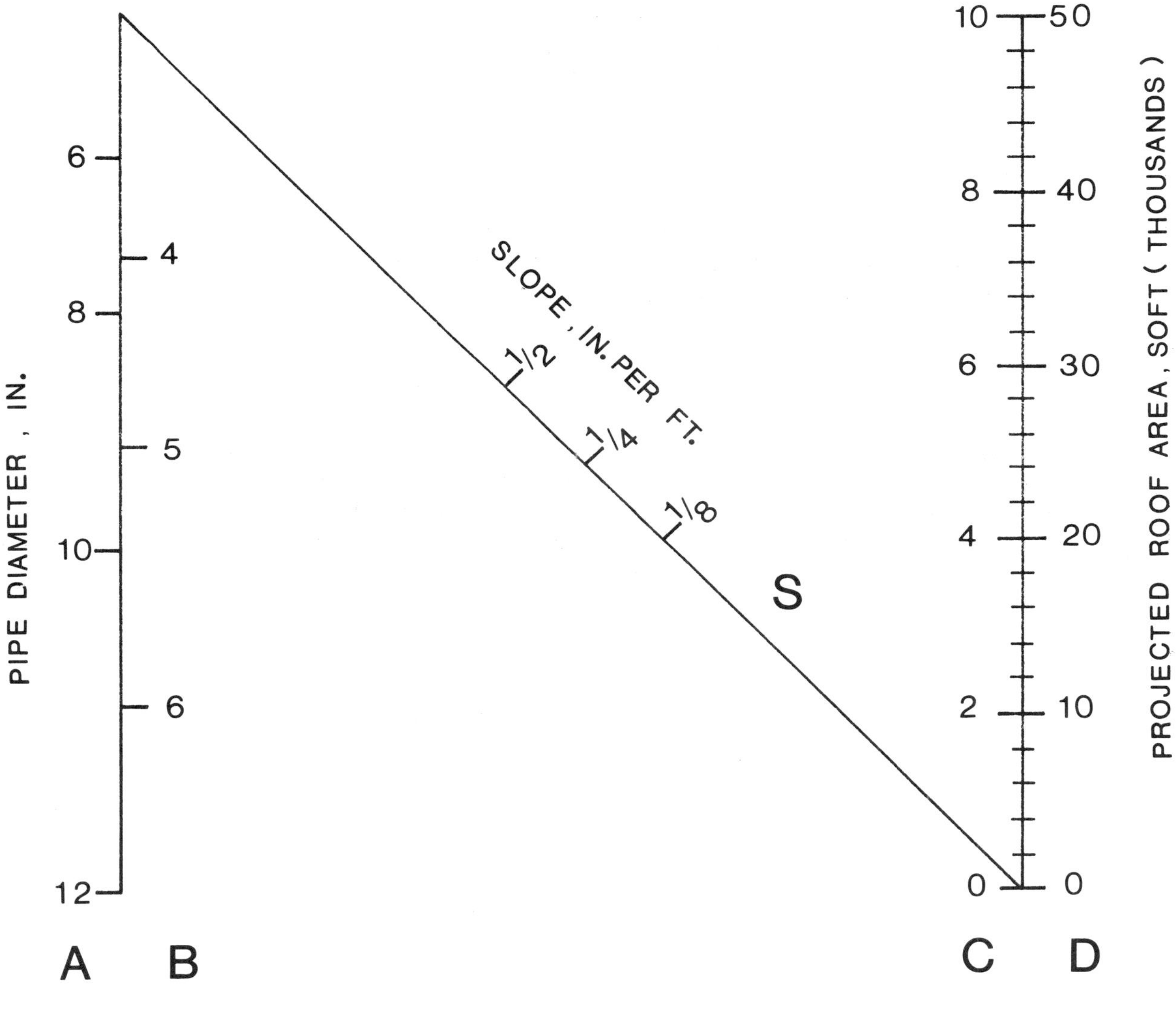
PIPE DIAMETER , IN.
6
8
10
12
4
5
6
A
B
SLOPE , IN. PER FT.
1/2
1/4
1/8
S
10
8
6
4
2
0
50
40
30
20
10
0
C
D
PROJECTED ROOF AREA , SOFT ( THOUSANDS )

## 2–14. Determines trench excavation quantities

This nomograph can be used to determine the quantity of material in cubic yards of excavation per lineal ft of trench for direct burial of pipes, pipe tunnels, footings, etc. The determination of the quantity of excavated material in cubic yards is required to calculate the cost to excavate and haul the material; these costs are usually expressed per cubic yard.

The number of cubic yards of material is obtained from the formula:

$$CY/LF = \tfrac{1}{2}(B + T)D/27$$

where
- $CY/LF$ = quantity of material, $yd^3$/lineal ft
- $B$ = bottom width of the trench, ft
- $T$ = top width of the trench, ft
- $D$ = depth of the trench, ft

### EXAMPLE

Determine the quantity of excavated material in cubic yards per lineal ft for a pipe trench where the size of the excavation is 10 ft deep by 6 ft wide.

### Solution

Align 10 ft on the D Scale with 6 ft on the W scale. Read the answer on the M scale as 2.22 $yd^3$/lineal ft. Note: For trenches with dimensions beyond the range of the nomograph, use the same multiplier for the M scale as used on the D or W scales. For example, if the trench is to be 10 ft deep by 12 ft wide, use 2 × 6 ft and multiply the answer, 2.22 $yd^3$/lineal ft by 2 and obtain 4.44 as the solution.

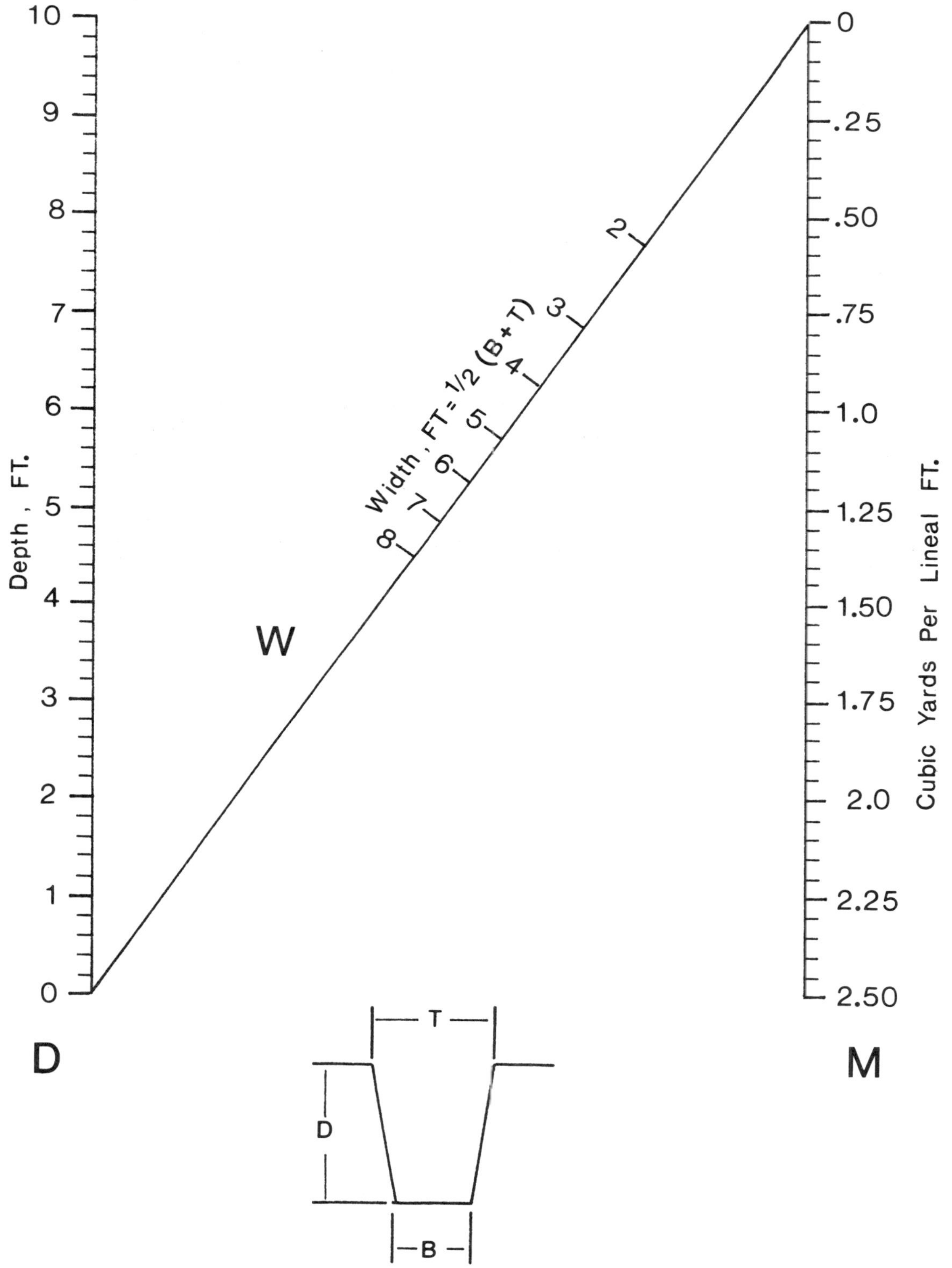
10
9
8
7
6
5
4
3
2
1
0
Depth , FT.
D
W
Width , FT = 1/2 (B+T)
2
3
4
5
6
7
8
0
.25
.50
.75
1.0
1.25
1.50
1.75
2.0
2.25
2.50
Cubic Yards Per Lineal FT.
M
T
D
B

## 2–15. Universal concrete volume calculator

This nomograph can be used to quickly determine the quantity of concrete, in cubic yards, required for slab on grade, foundation wall, retaining walls, etc., where the dimensions are in feet and inches. No need to worry about the formula or the calculation, just align the dimensions of the problem and read the answer, in cubic yards.

### EXAMPLE

Determine the volume of concrete required for a foundation 25 feet long, 10 feet high and 12 inches thick.

### Solution

Align 25 ft on scale L with 10 ft on scale W and mark the intersection on scale A (this point is the surface area of the foundation wall, 250 $ft^2$). Align this mark on scale A with 12 in. on scale T and read 9.25 $yd^3$. Note: For problems with dimensions beyond the range of the scales, use the same multiplier on scale V as was used on any *one* of the other scales. For example, for a 50-ft (2 × 25) wall above, the answer will be 18.5 $yd^3$ (2 × 9.25).

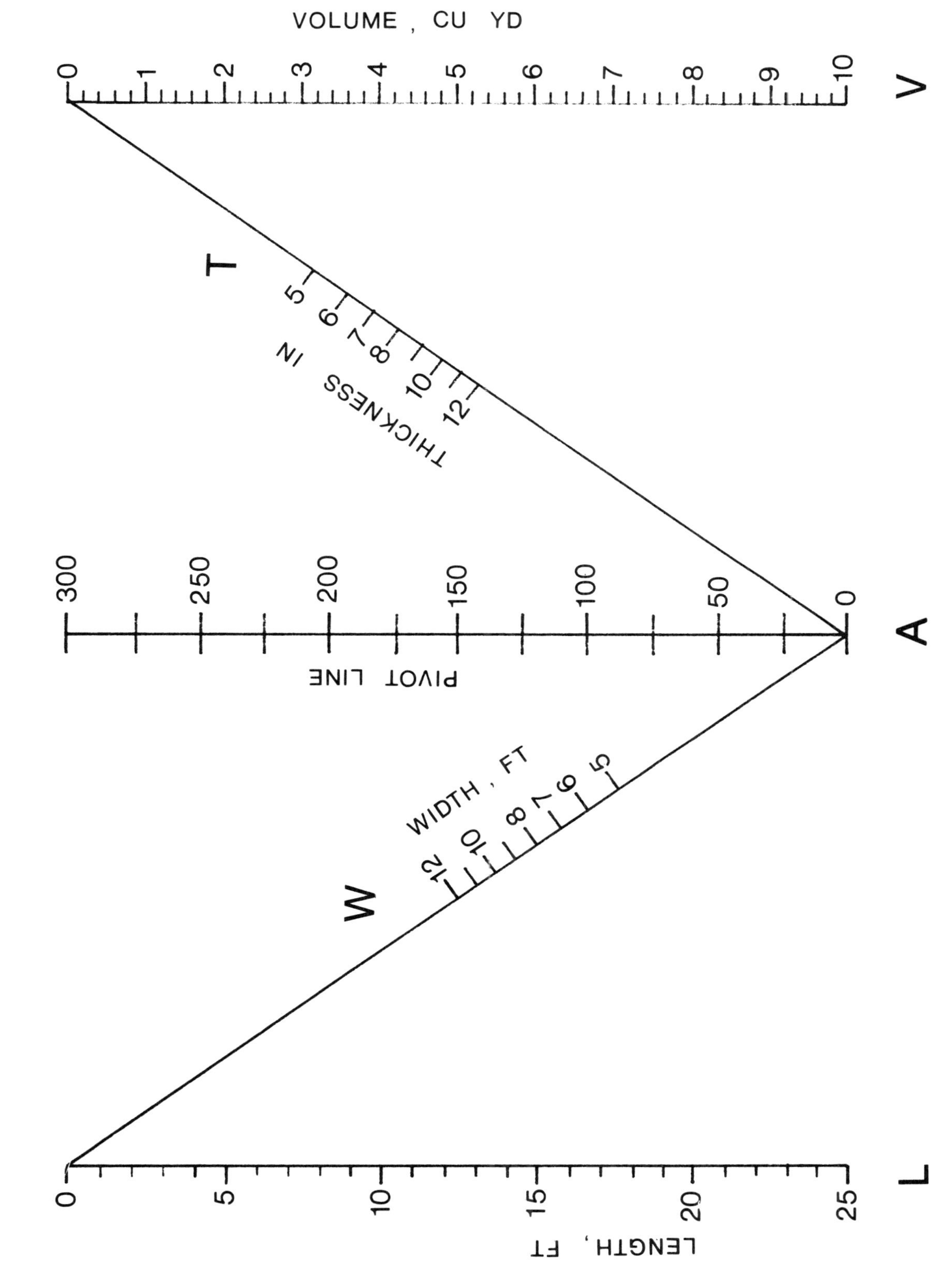
VOLUME , CU YD
V
0 1 2 3 4 5 6 7 8 9 10
T
THICKNESS IN
5 6 7 8 10 12
PIVOT LINE
A
300 250 200 150 100 50 0
W
WIDTH , FT
12 10 8 7 6 5
L
LENGTH, FT
0 5 10 15 20 25

# GROUP 3

# ELECTRICAL, MECHANICAL AND COSTS

## 3–1. Determining power loss in conductors

Here is a handy nomograph to determine quickly and accurately the power loss in insulated copper electrical conductors in sizes from No. 14 to 500 MCM in metal raceway when carrying allowable ampacities. It is based on data given in Table 310–16 (Article 310) and Tables 8 and 9 (Chapter 9) of the National Electric Code. It provides watt losses for 2-wire (single-phase) and 3-wire (3-phase) circuits for runs up to 250 ft. long.

### EXAMPLE 1

Determine the power loss in a 2-wire circuit 200-ft long with No. 6 conductors carrying their rated current of 65 A.

### Solution

Using a straightedge, align 200 ft on scale L with size No. 6 on scale S and read 693 W on scale W2.

### EXAMPLE 2

Same conditions as in first example, except that a 3-wire circuit is involved.

### Solution

Align straightedge as above but read 1040 W on scale W3. Note: On runs beyond the range of scale L, divide the actual length of the run by a number that yields a value within the range of scale L. Determine the power loss as above, then multiply that answer by the same factor used in the division. The result will be the actual power loss.

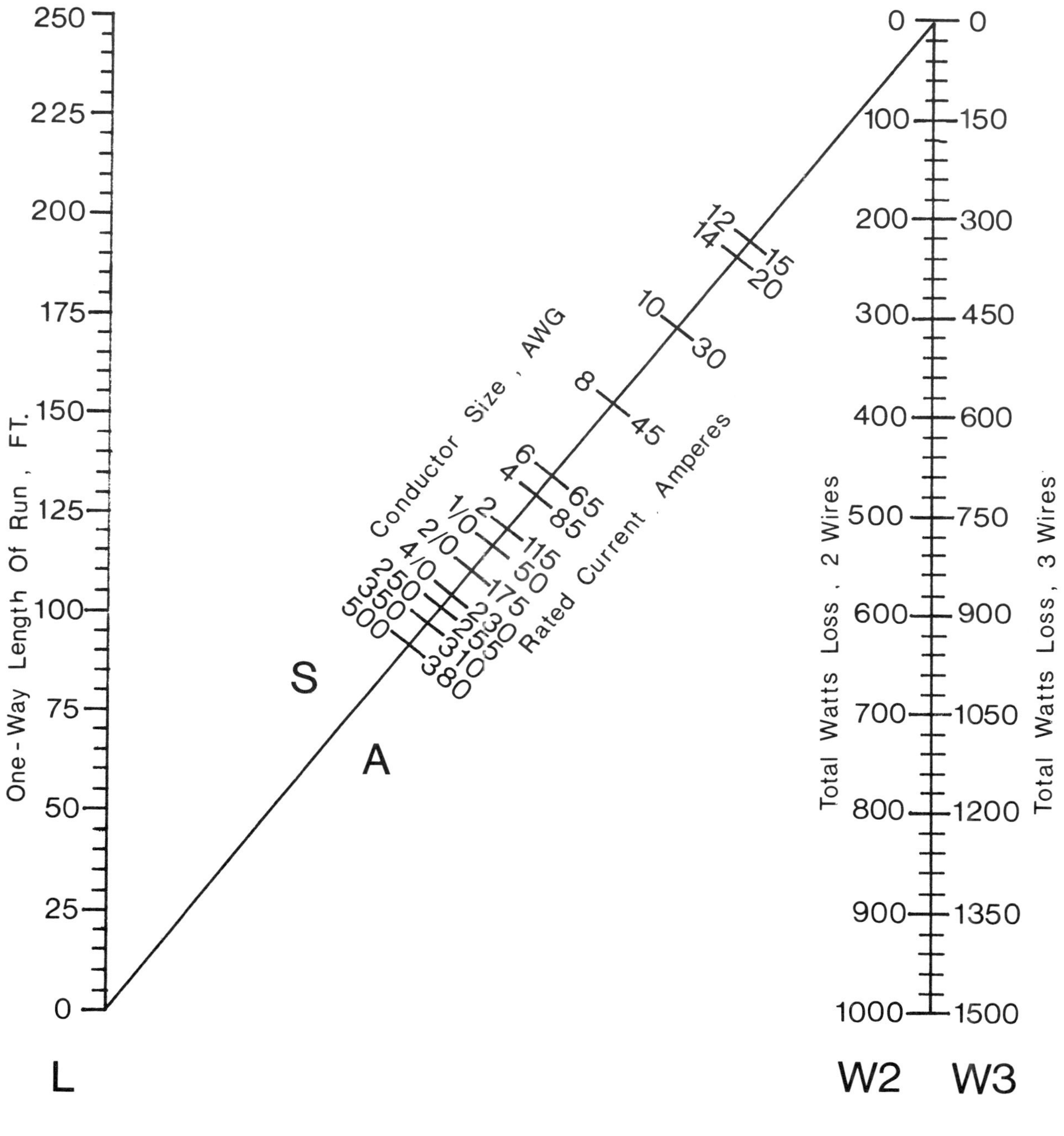

One-Way Length Of Run, FT.
250
225
200
175
150
125
100
75
50
25
0
L
S
A
Conductor Size, AWG
12
14
10
8
6
4
2
1/0
2/0
4/0
250
350
500
Rated Current, Amperes
15
20
30
45
65
85
115
150
175
230
255
310
380
Total Watts Loss, 2 Wires
Total Watts Loss, 3 Wires
0
100
200
300
400
500
600
700
800
900
1000
0
150
300
450
600
750
900
1050
1200
1350
1500
W2
W3

## 3–2. Determining current-kVA relationships

The nomograph provides a quick graphical solution for determining the relationship between current and kVA in both single- and three-phase power systems rated at a nominal 120, 208, 240 or 480 V.

### EXAMPLE 1

How much transformer capacity will be required to serve a three-phase, 208 V load drawing 50 A?

### Solution

Align 50 A on scale E with 208 V on scale C, extend the line to scale A, and read transformer capacity as 18 kVA.

### EXAMPLE 2

Can this additional load be assumed by a 100-kVA transformer presently carrying 220 A?

### Solution

Align 100 kVA on scale A with 208 V on scale C, extend the line to scale E, and read the current-carrying capacity of the 100-kVa transformer as 277 A. The transformer can, therefore, assume the additional load without overloading.

The nomograph is direct-reading, but can be used for larger or smaller values of current and kVA than those shown. For example, on a three-phase circuit, a 500-A (2 × 250 on scale E), 208-v load would require 180 kVA (2 × 90 on scale A), and a 5 A, 208-V load would require 1.8 kVA.

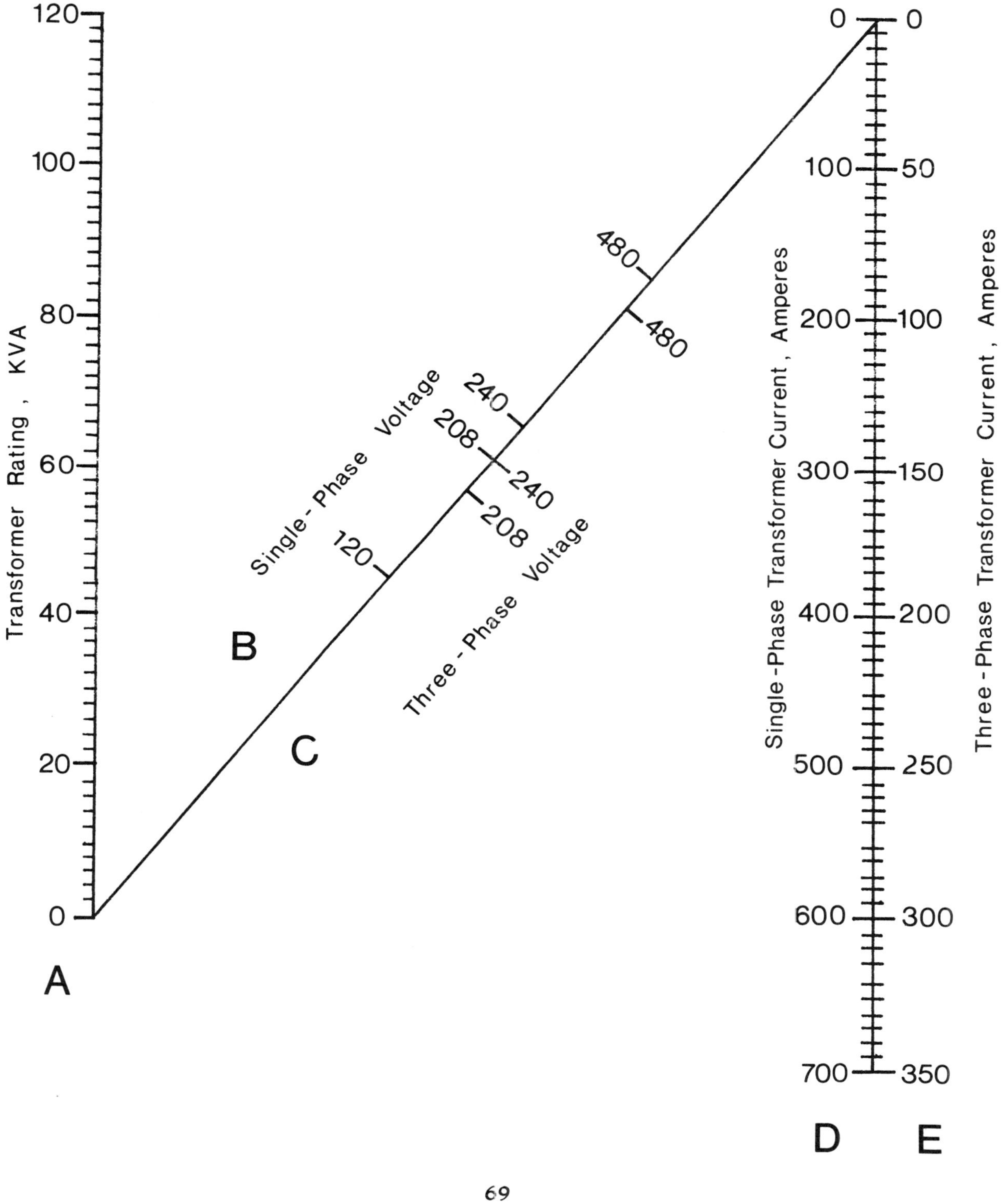

Transformer Rating, KVA
120
100
80
60
40
20
0
A
Single - Phase Voltage
Three - Phase Voltage
120
208
240
480
B
C
Single-Phase Transformer Current, Amperes
Three - Phase Transformer Current, Amperes
0
100
200
300
400
500
600
700
50
150
250
350
D
E

## 3–3. Determining current kilowatt relationships for resistance loads

The nomograph provides a quick graphical solution for determining the relationship between current and kW in both single- and three-phase power systems rated at a nominal 120, 208, 240 and 480 V.

### EXAMPLE

Determine the current that will flow in a single-phase circuit of 120 V with a 6-kW resistance load.

### Solution

Align 6 kW on scale A with 120 V on scale B and read 50 A on scale D.

The nomograph is direct-reading, but can be used for larger or smaller values of current and kW than those shown. For example, on a single-phase circuit as in the above example, the resistance load is now 12 kW (use 2 × 6 on scale A), and read (2 × 50 on scale D) 100 A.

The same multiplier must be used on kW and current scales to solve problems beyond the range of the scales as shown.

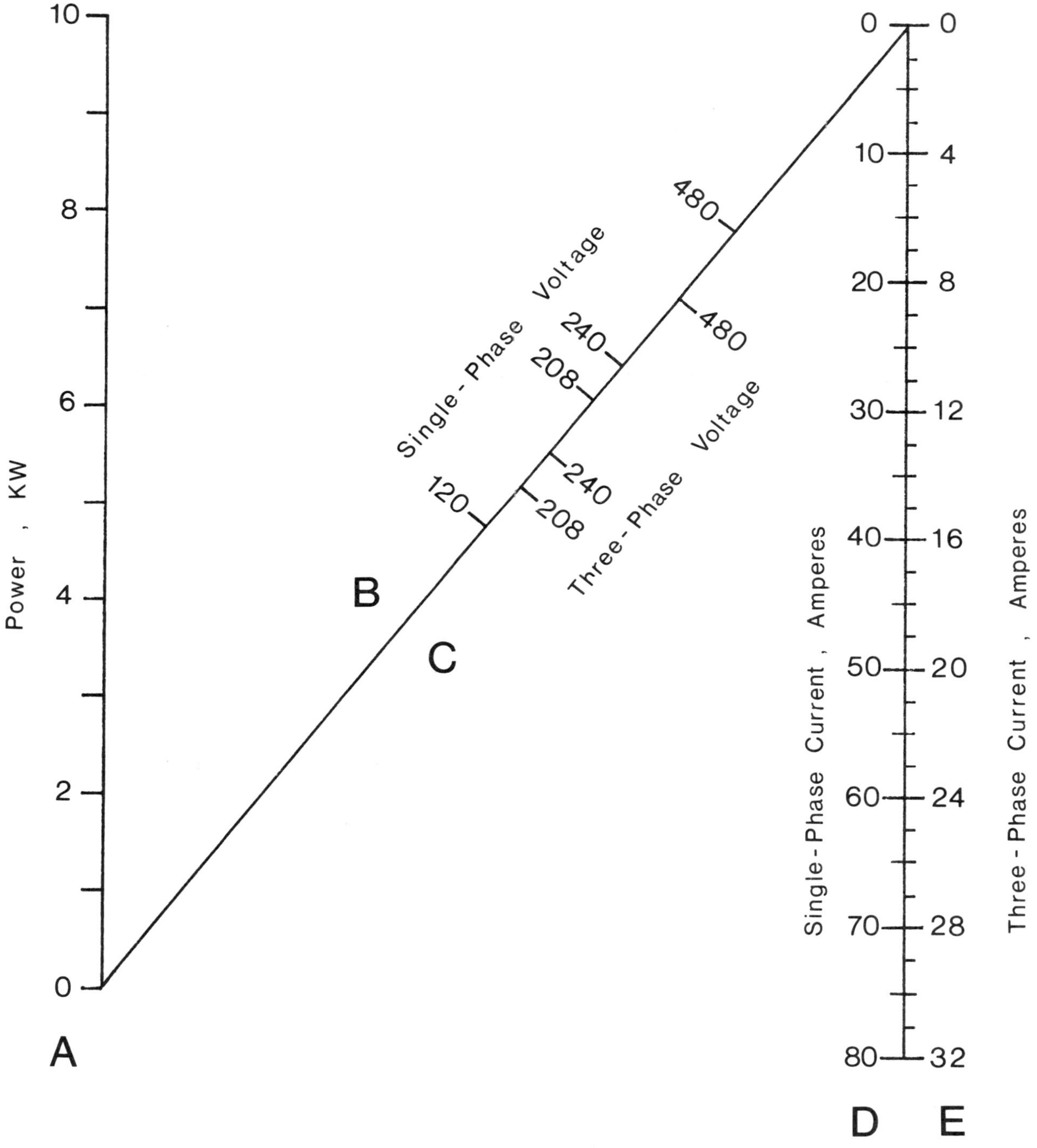

Power , KW
10
8
6
4
2
0
A
Single - Phase Voltage
120
208
240
480
Three - Phase Voltage
208
240
480
B
C
Single - Phase Current , Amperes
0
10
20
30
40
50
60
70
80
Three - Phase Current , Amperes
0
4
8
12
16
20
24
28
32
D
E

## 3–4. Ohm's law calculator

This nomograph is useful in solving Ohm's law and power problems, and is from the formulas:

$$\text{Ohm's law: } I = E/R$$

$$\text{power: } P = EI$$

where
- I = current, A
- E = single phase voltage, V
- R = resistance, ohms
- P = power, kW

**EXAMPLE**

Determine the current that will flow in a 120-V circuit when the resistance load is 2.4 kW. Determine the resistance of the electric heater.

**Solution**

Align 2.4 on scale P with 120 on scale $V_1$ and read 20 A on scale I. Next align 20 on scale I with 120 on scale $V_2$ and read 6 Ω on scale O. Note: Scales P, I and O, and $P_1$, $I_1$ and $O_1$ are companion scales and must be used together to solve the problems.

Problems can be solved when any two of the four variables are known. For example, suppose that the problem were presented in the following form: determine the current and power of a 120-V circuit when the resistance of the electric heater is 6 Ω.

Align 120 on scale $V_2$ with 6 on scale O and read 20 A on scale I. Then align 20 on scale I with 120 on scale $V_1$ and read 2.4 kW on scale P.

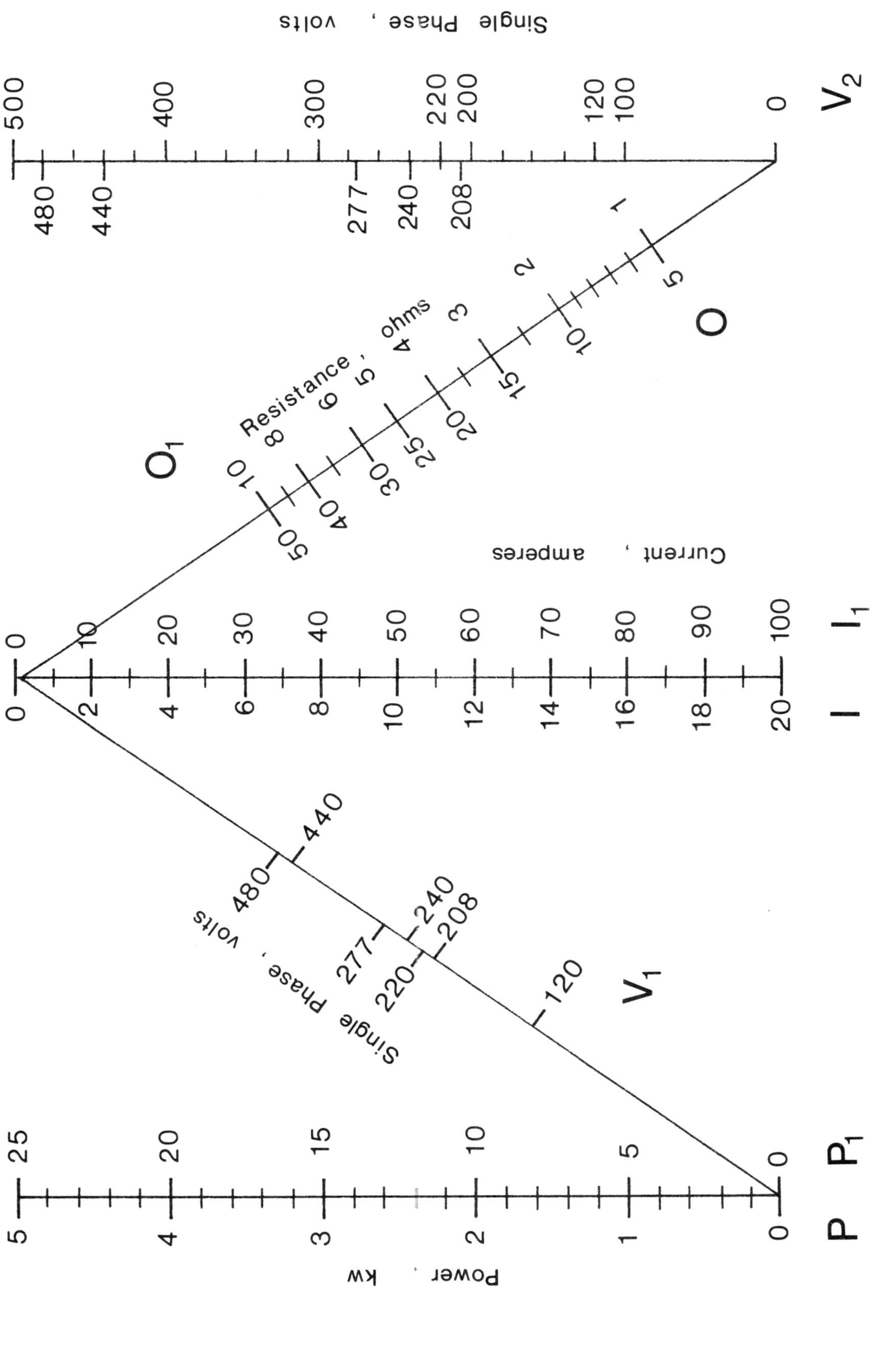
Power, kw
P
P1
Single Phase, volts
V1
120
208
220
240
277
440
480
I
I1
Current, amperes
O1
O
Resistance, ohms
V2
0
100
120
200
208
220
240
277
300
400
440
480
500

## 3–5. Parallel resistance calculator

This nomograph is useful in determining the equivalent resistance value in ohms when resistors are connected in parallel. The equivalent value is obtained from the formula:

$$R_e = \frac{R_1 \times R_2}{R_1 + R_2}$$

### EXAMPLE

Determine the equivalent value of two resistors connected in parallel that have values of 15 and 10 ohms.

### Solution

Align 15 on scale $R_1$ with 10 on scale $R_2$ and read 6 on scale $R_e$. Note: When more than two resistors are connected in parallel, first solve for any two, then solve for the third, and so on.

When the value of the resistor is different from the range of the scales, the values of both scales can be divided or multiplied by 10, 100, 1000, etc., as required.

For example, suppose the value in the problem were 1.5 and 1.0 ohms. Divide all scales by 10, and the answer is now 0.6 ohms.

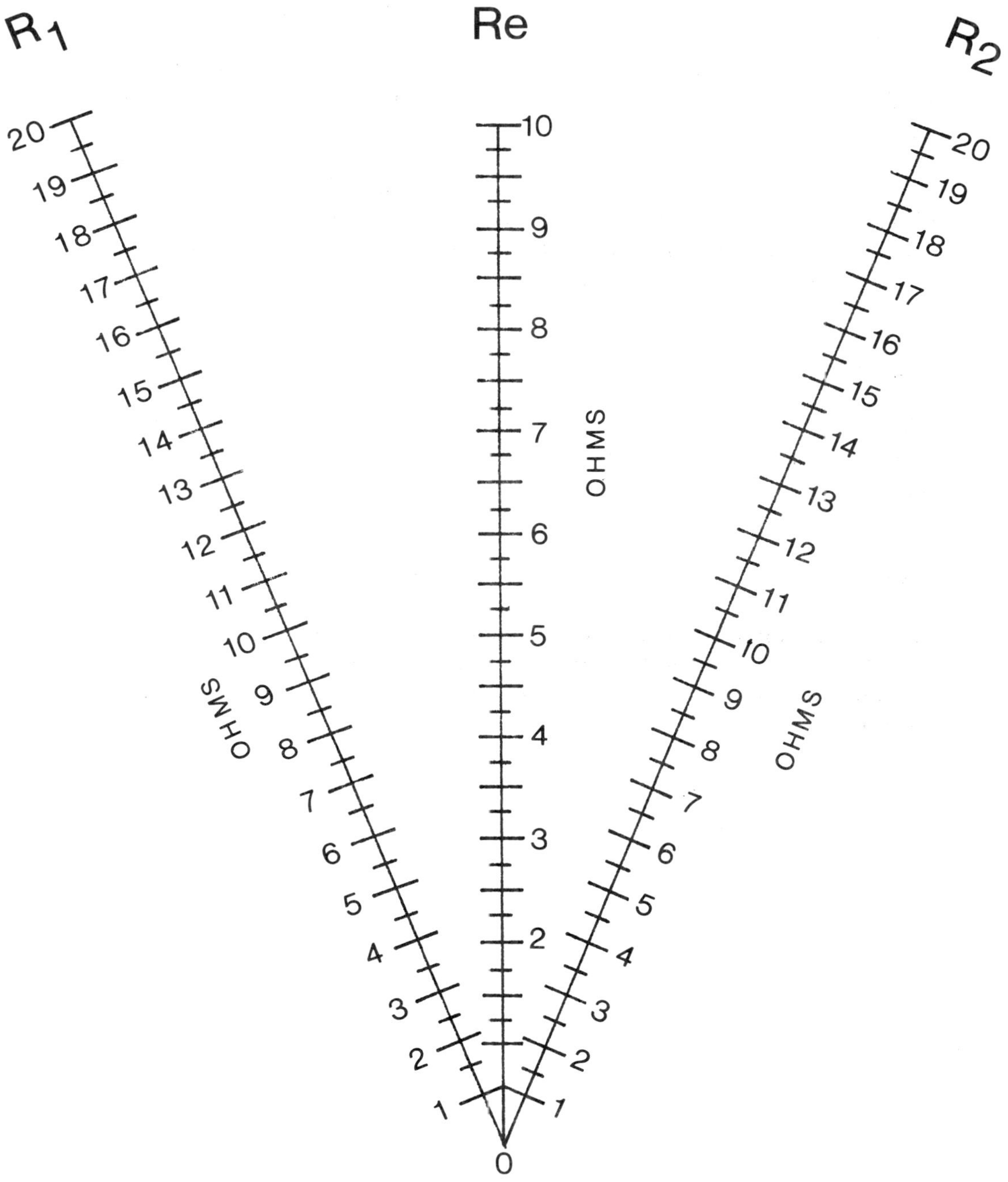
R1
Re
R2
OHMS
OHMS
OHMS
20
19
18
17
16
15
14
13
12
11
10
9
8
7
6
5
4
3
2
1
0

## 3-6. Determining cost savings for equipment shut downs

Do you know how much money you spend unnecessarily or how much energy you waste permitting electric-motor–driven equipment to operate during unoccupied time at your facility?

With this nomograph you can quickly and easily determine the dollars saved annually when electric motors are shut off during unoccupied times.

This nomograph was constructed for occupied time of 10 hours per day, 5 days per week and observing major holidays. Time other than the defined occupied time is unoccupied time. The motors used for the nomograph are standard three-phase, totally enclosed, fan-cooled, 60 Hz, 1750 rpm, operating on 460 V.

### EXAMPLE

Determine the annual saving in dollars when a 5-hp circulating pump on pumped hot water circulating system is shut off during the unoccupied time and the average power cost is 4¢/kWh.

### Solution

Align 5 hp on scale A with 4¢ on scale C and read $950 annual saving on scale E. Note: Scales A, C and E are companion scales for 0- to 5-hp motors, and scales B, D and F are companion scales for 0- to 25-hp motors.

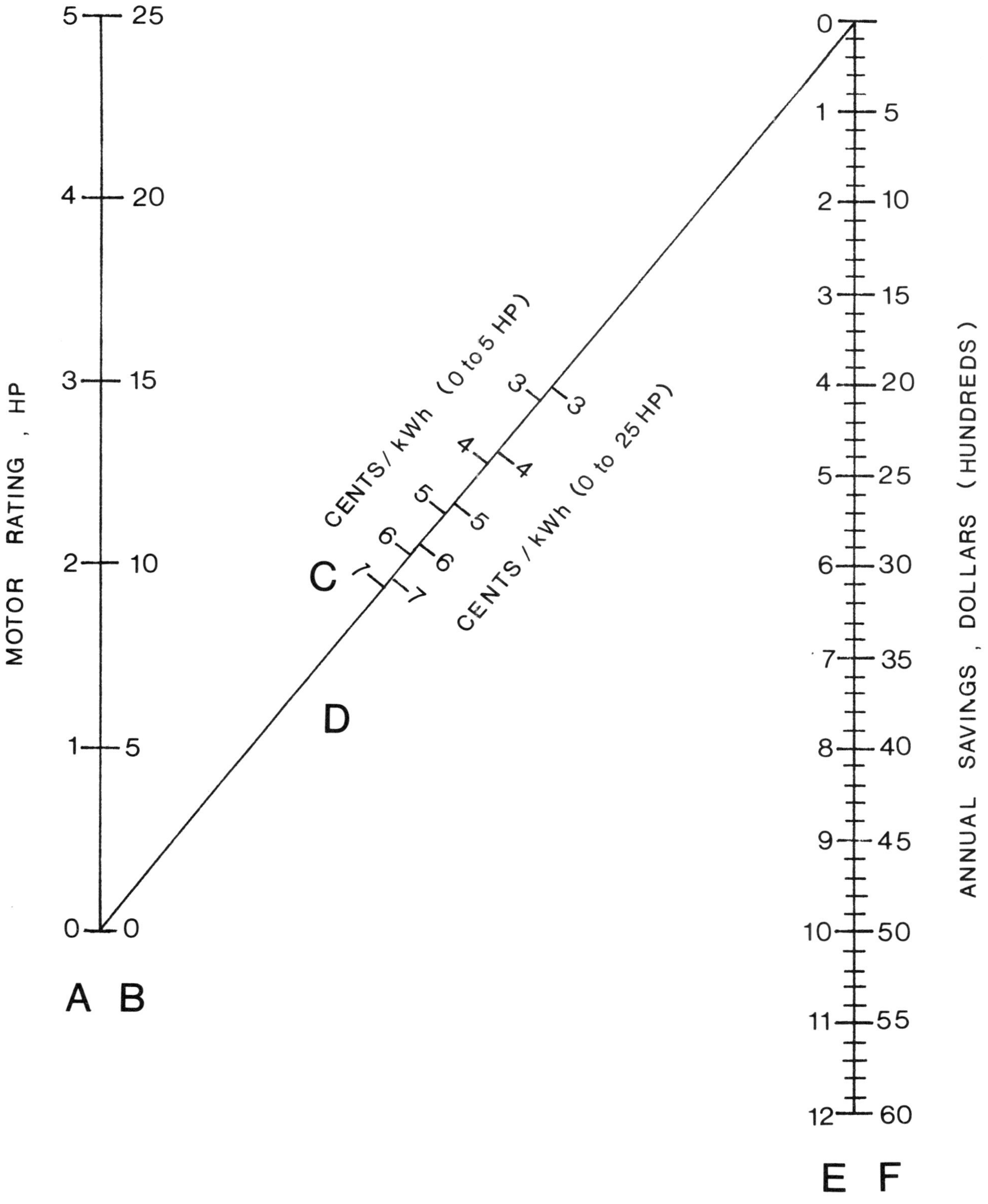

MOTOR RATING , HP
5
25
4
20
3
15
2
10
1
5
0
0
A B
CENTS / kWh (0 to 5 HP)
CENTS / kWh (0 to 25 HP)
3
4
5
6
7
C
D
ANNUAL SAVINGS , DOLLARS (HUNDREDS)
0
1
5
2
10
3
15
4
20
5
25
6
30
7
35
8
40
9
45
10
50
11
55
12
60
E F

## 3–7. Finding belt velocity

Ever have to know the velocity at which a belt of any kind V, flat, timing or conveyor is moving? How about the size of pulley needed to achieve a specified belt velocity? With this handy nomograph, belt velocity, pulley diameter or pulley speed can be determined quickly when the other two quantities are known.

The relationship between belt velocity and pulley speed and diameter is defined by the equation:

$$V = \pi D/12 \times S$$

where
$V$ = belt velocity, fpm
$D$ = pulley diameter, in.
$S$ = pulley speed, rpm

### EXAMPLE

Determine the speed of the head pulley that will drive a conveyor belt at 400 fpm, if the head pulley is 12 in. in pitch diameter.

### Solution

On the nomograph, align 400 fpm on scale V with 12 in. on scale D. Extend the line to scale S and read the head pulley speed as 127 rpm.

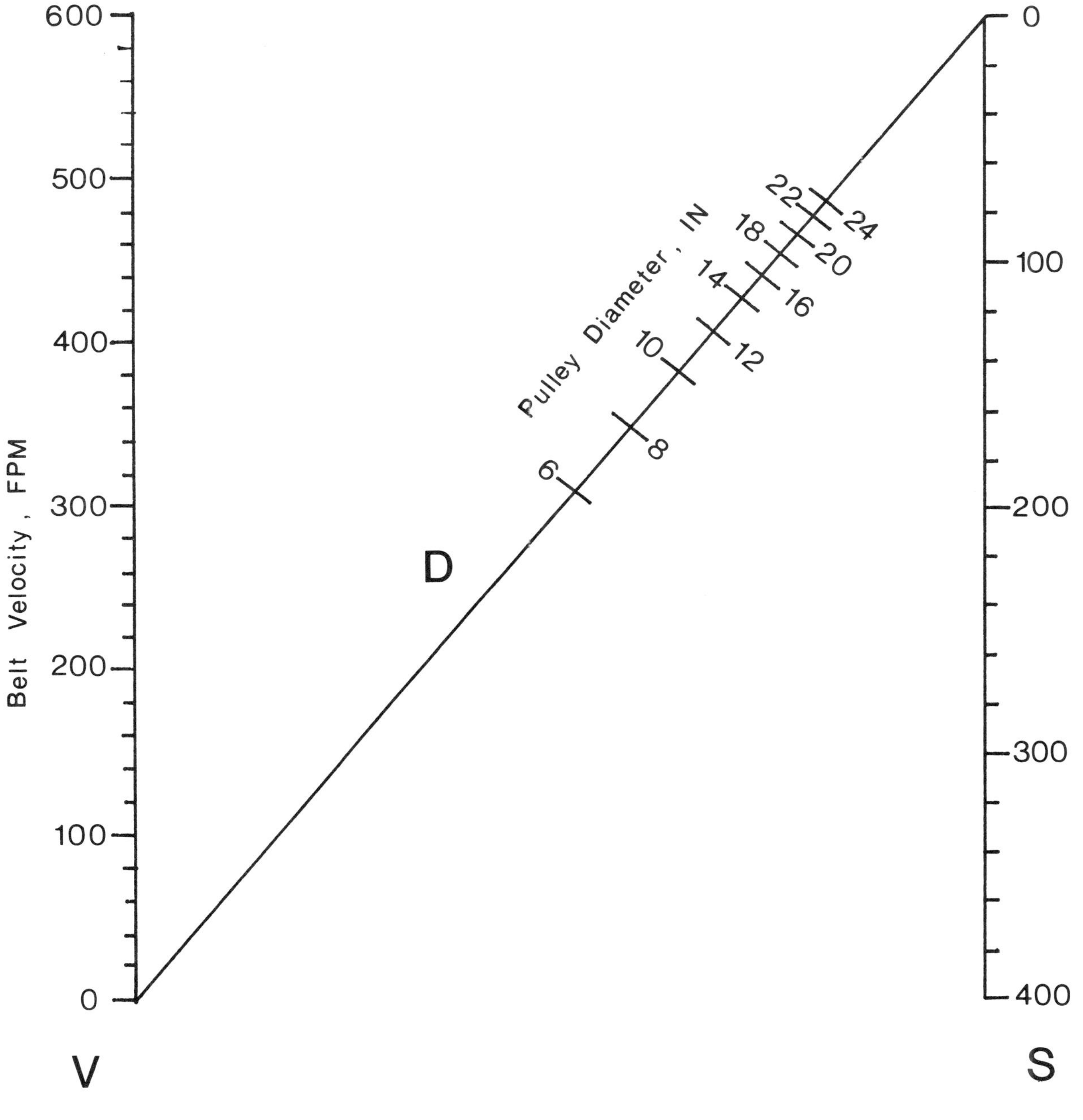
600
500
400
300
200
100
0
Belt Velocity , FPM
V
D
Pulley Diameter , IN
6
8
10
12
14
16
18
20
22
24
0
100
200
300
400
Pulley Speed, RPM
S

## 3–8. Sizing drive pulleys

This handy nomograph can be used to determine the speed of fans, pumps, generators and similar equipment, in revolutions per minute, when the equipment is belt-connected and driven by an electric motor with a synchronous speed of either 1200 or 1800 rpm. It can also be used to determine the correct pulley sizes when the desired speed of the driven equipment is known.

The pulley-size-to-speed relationship is defined by the equation:

$$N \times D = n \times d$$

where
$N$ = driven pulley speed, rpm
$n$ = drive pulley speed, rpm
$D$ = driven pulley pitch diameter, in.
$d$ = drive pulley pitch diameter, in.

The nomograph is based on full load speeds of 1150 and 1750 rpm. It is good practice to select one adjustable pulley, so that exact design speed can be achieved. For the sake of economy, the smaller pulley should be the adjustable one.

### EXAMPLE

When the drive is to be a standard 1800-rpm, three-phase induction motor, find the pitch diameters of the pulleys required to power an exhaust fan at 900 rpm.

#### Solution

On the nomograph, align 900 on scale N with an appropriate exact size on scale D—in this case, 12 in. Extend the line to scale d and read the drive pulley size as 6.2 in.

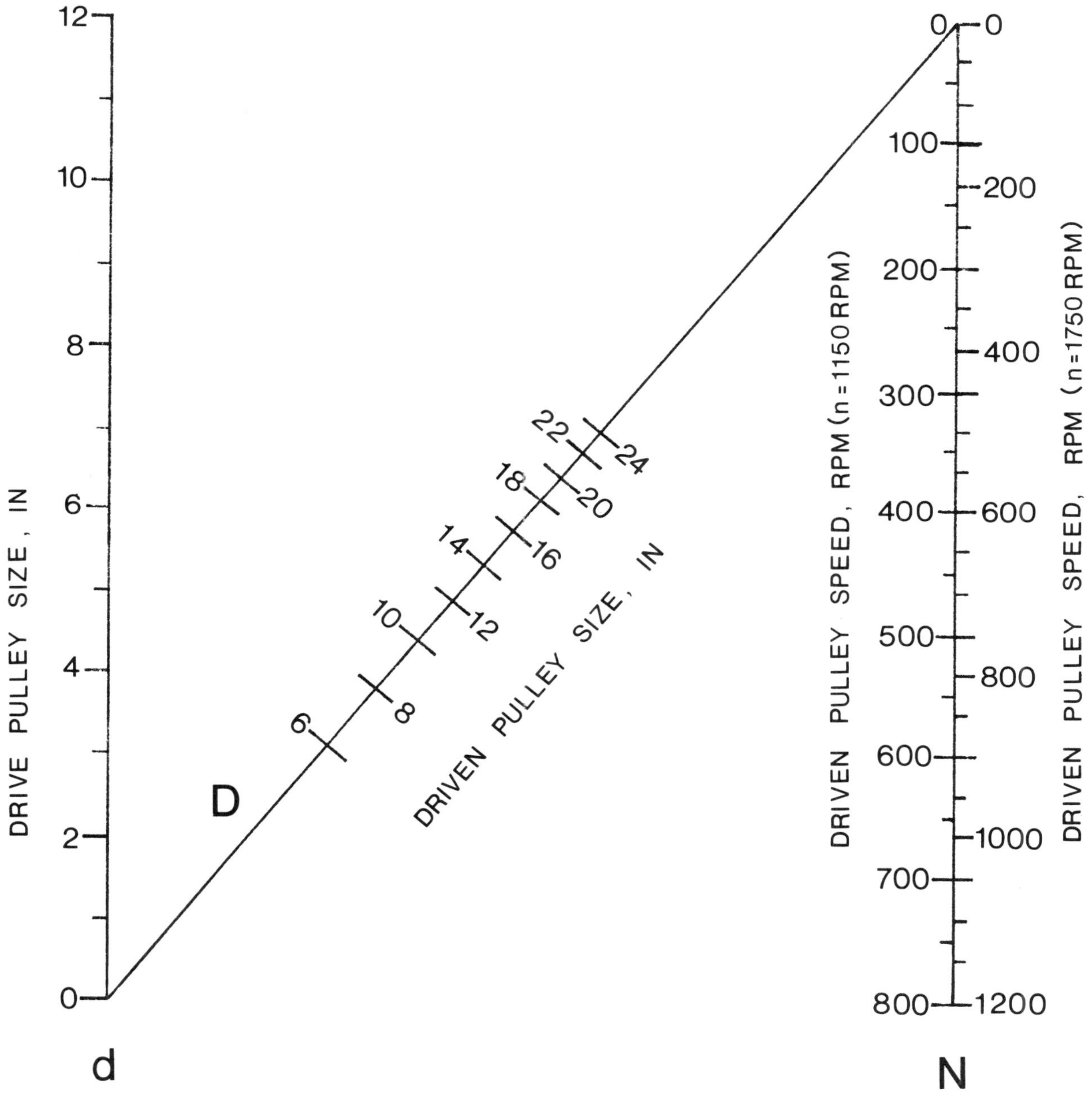
DRIVE PULLEY SIZE, IN
d
0
2
4
6
8
10
12
D
DRIVEN PULLEY SIZE, IN
6
8
10
12
14
16
18
20
22
24
N
DRIVEN PULLEY SPEED, RPM (n=1150 RPM)
0
100
200
300
400
500
600
700
800
DRIVEN PULLEY SPEED, RPM (n=1750 RPM)
0
200
400
600
800
1000
1200

## 3-9. Duct weight calculator

When designing a ventilation system for makeup air or pollution exhaust, it is usually necessary to calculate the total number of pounds of material required to determine the cost for the system. The nomograph provides a rapid and easy method for determining the weight per linear foot of galvanized sheet steel.

To calculate weight per linear foot, select either the diameter for round duct or the perimeter of rectangular duct on the A scale. On scale B select the required gauge for the sheet steel of the duct. Draw a line from A to B extending it to scale C. Read the duct weight in pounds per linear foot on C.

When duct sizes are beyond the range of scale A, enter one-half the diameter of the round duct or one-half the perimeter of the rectangular duct on scale A. It will then be necessary to double the answer obtained on scale C.

### EXAMPLE

Find the total weight of a 25-in. round duct 35 ft long. Duct material is 24 gauge sheet steel.

#### Solution

Align 25 on the round duct portion of scale A with 24 on scale B, and read 7.5 lb/lin ft on scale C. Multiply 7.5 by 35, and find the weight of the duct to be 263 lb.

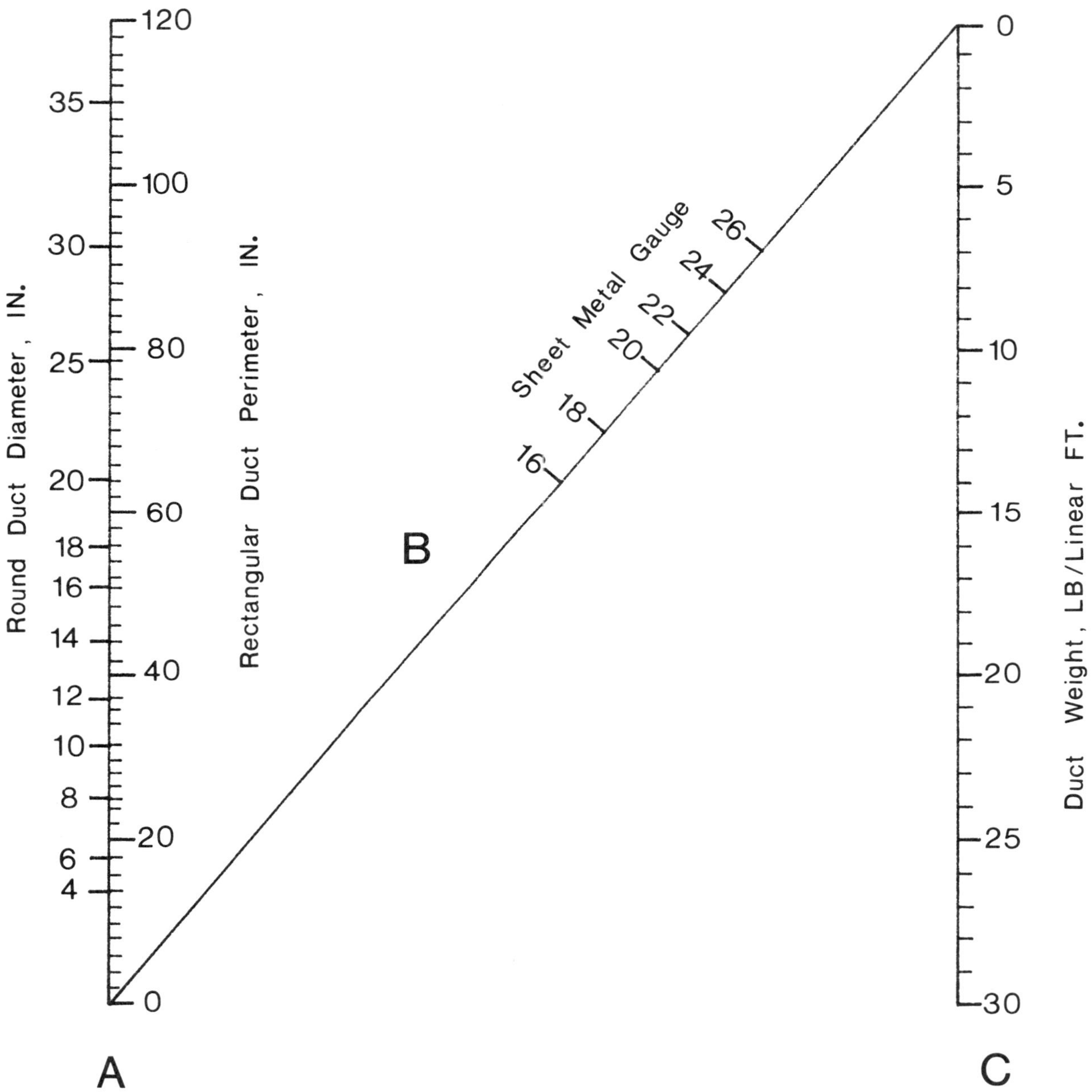
Round Duct Diameter, IN.
Rectangular Duct Perimeter, IN.
Sheet Metal Gauge
Duct Weight, LB/Linear FT.
A
B
C
120
100
80
60
40
20
0
35
30
25
20
18
16
14
12
10
8
6
4
16
18
20
22
24
26
0
5
10
15
20
25
30

## 3–10. Estimating sheet-metal duct costs

This nomograph can be used to estimate the cost of sheet-metal duct for HVAC systems. The plant engineer selects the size of round or rectangular duct and the gauge of sheet metal needed to handle the job. Then, plotting a single line on the nomograph gives the estimated cost per linear foot of duct.

The cost is based on $2/lb galvanized sheet metal in 500- to 1000-lb quantities. For amounts less than 500 lb, add 10% to the cost; for quantities more than 1000 lb, deduct 10% from the cost.

### EXAMPLE

What is the estimated cost of 30 ft of 25-in. round duct constructed from 24 gauge galvanized sheet metal?

### Solution

On the nomograph, connect 25 in. on scale A with 24 gauge on scale B, and extend the line to scale C. Read the cost per linear foot as approximately $15 or $450 for the 30-ft section.

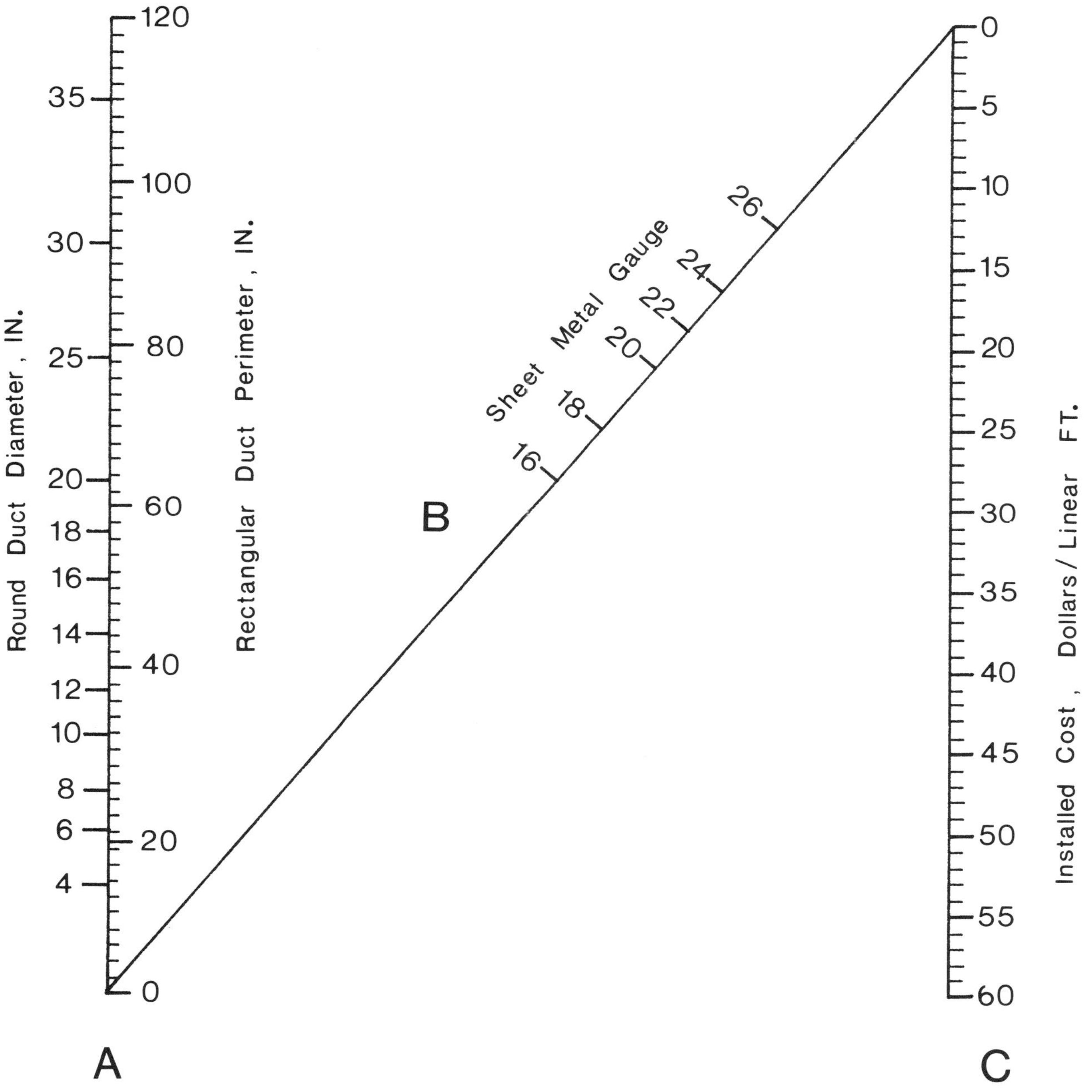

Round Duct Diameter, IN.
Rectangular Duct Perimeter, IN.
35
30
25
20
18
16
14
12
10
8
6
4
120
100
80
60
40
20
0
A
B
Sheet Metal Gauge
16
18
20
22
24
26
C
Installed Cost, Dollars/Linear FT.
0
5
10
15
20
25
30
35
40
45
50
55
60

## 3-11. Calculating sound levels for fan speed changes

One of the most common sources of in-plant noise, particularly in warm weather, is fans. Since fan speeds are frequently altered to effect changes in delivered air quantity and static pressure, the plant engineer needs to be able to determine what the consequent noise effects will be.

The commonly employed unit for measuring relative sound power level is the decibel (dB) which by definition is 10 times the common (base 10) logarithm of the ratio of the noise pressure to 0.0002 $\mu$bar.

If the noise level of a single fan at a particular point of rating is known, the change in decibels (dB) resulting from the variation in speed can be estimated from:

$$\text{dB} = 50 \log_{10} \left[ \frac{(\text{rpm})_2}{(\text{rpm})_1} \right]$$

where $(\text{rpm})_1$ is the original speed, and $(\text{rpm})_2$ is the new speed.

A quick graphical solution to this equation is provided by nomograph 3-11.

### EXAMPLE

What is the sound level change if a fan speed is increased from 1160 to 1750 rpm?

### Solution

On nomograph 3-11, connect 1160 on the $(\text{rpm})_1$ scale with 1750 on the $(\text{rpm})_2$ scale. At the point where this line intersects the dB scale, read the change in sound level as an increase of 9 dB.

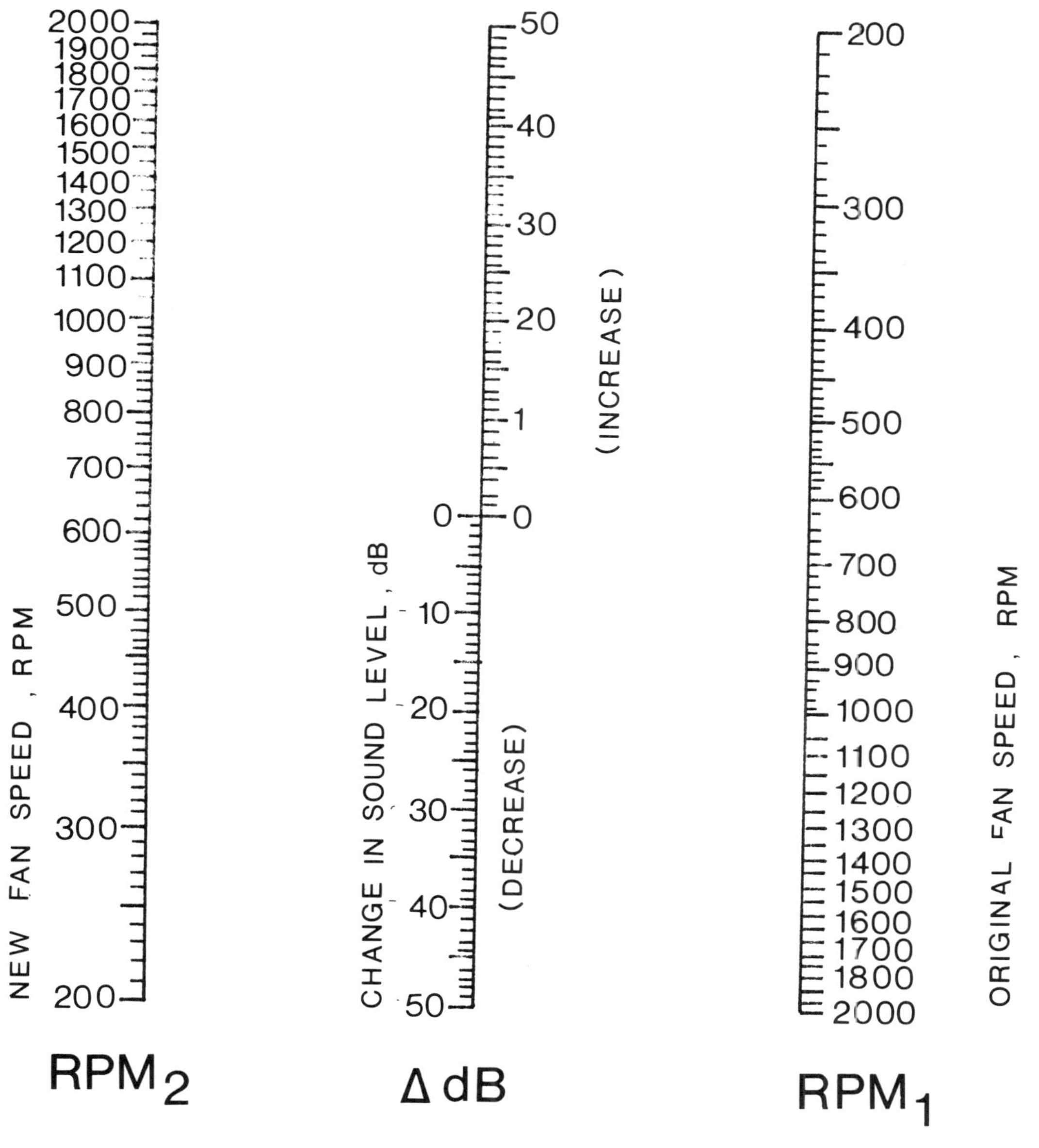
NEW FAN SPEED , RPM
2000
1900
1800
1700
1600
1500
1400
1300
1200
1100
1000
900
800
700
600
500
400
300
200
RPM2
CHANGE IN SOUND LEVEL , dB
(INCREASE)
(DECREASE)
50
40
30
20
1
0
10
20
30
40
50
Δ dB
ORIGINAL FAN SPEED , RPM
200
300
400
500
600
700
800
900
1000
1100
1200
1300
1400
1500
1600
1700
1800
2000
RPM1

## 3-12. Calculating sound levels for two or more fans

Because of the logarithmic relationship, the average sound level produced by several fans operating in the same area is not the sum of the individual sound levels. When several sources are contributing to the sound level, the average sound level, $(dB)_T$, can be determined from:

$$(dB)_T = 10 \log \left[ \text{antilog} \frac{(dB)_A}{10} + \text{antilog} \frac{(dB)_B}{10} + \text{antilog} \frac{(dB)_C}{10} + \ldots \right]$$

where subscripts A, B, C . . . indicate the individual sound sources.

Nomograph 3-12 permits a rapid solution of this equation. For sound level ranges not on the scales, values in multiples of 10 can be simultaneously added to or subtracted from all scale values. For more than two sound sources, the solution must be carried out in steps. First, solve for two sources. Let this answer be $(dB)_A$ and the sound level of the third source $(dB)_B$. Solve again, using the nomograph to find $(dB)_T$ for the three sources. Repeat this process for all the remaining sources.

### EXAMPLE

What is the combined (average) sound level of two fans which have individual sound levels of 41 dB (Fan A) and 42 dB (Fan B)?

#### Solution

Since the sound levels of these fans do not appear on the scales, 10 must be subtracted from the nomograph scale values to permit use of nomograph 3-12.

Therefore, connect 51 on the $(dB)_B$ scale (51 - 10 = 41) with 52 on the $(dB)_A$ scale (52 - 10 = 42). Where this line intersects the $(dB)_T$ scale, read $(dB)_T$ = 54.53 dB. Subtracting 10 from this value, gives the combined average sound level as 44.53 dB. Keep in mind that in adjusting the scale values for problem solutions, all scales must be adjusted by the same amount.

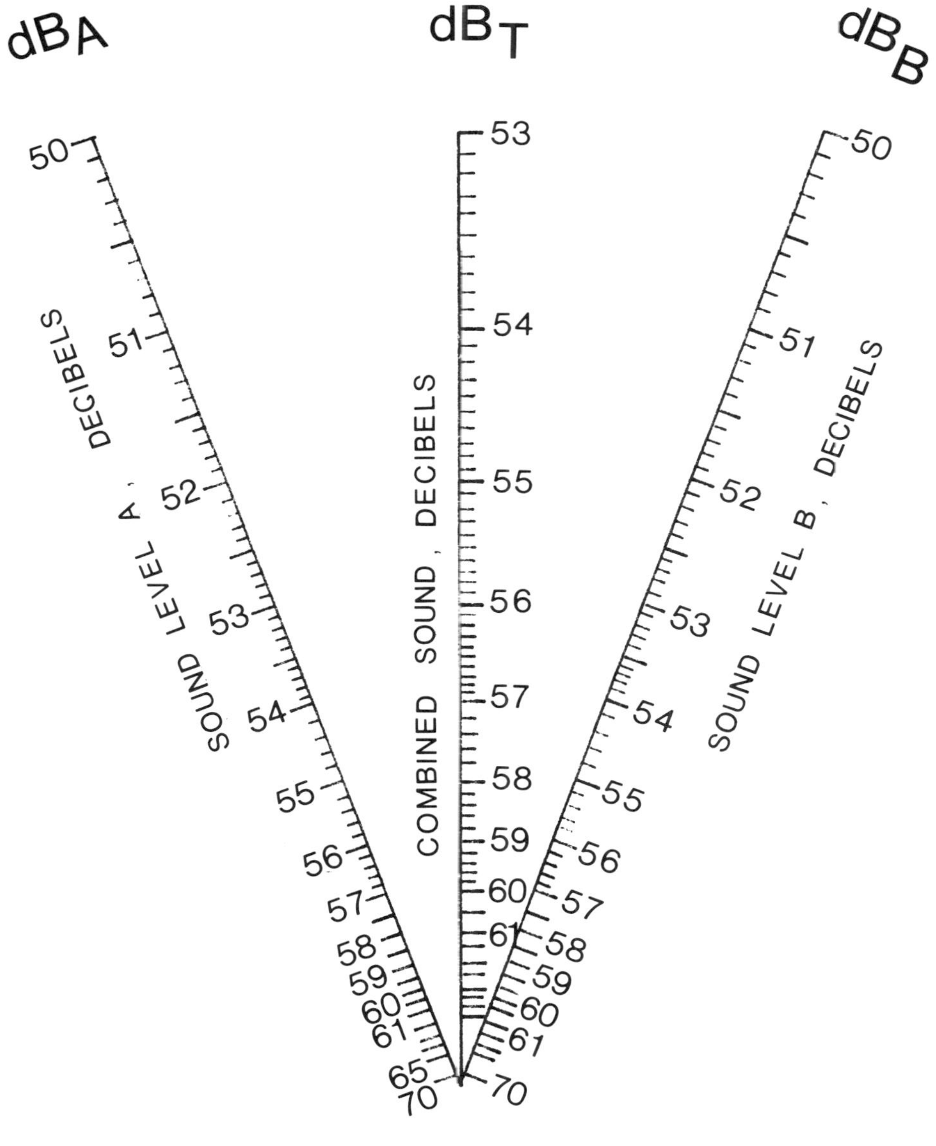
dBA
dBT
dBB
SOUND LEVEL A, DECIBELS
COMBINED SOUND, DECIBELS
SOUND LEVEL B, DECIBELS
50
51
52
53
54
55
56
57
58
59
60
61
65
70
53
54
55
56
57
58
59
60
61
50
51
52
53
54
55
56
57
58
59
60
61
70

## 3–13. Power factor calculator

This nomograph can be used to quickly solve electrical power problems involving kilowatts (kW), kilovolt-amperes (kVA), kilovolt-amperes reactive (kVAr), and power factor (pf) when any two of the parameters are known and for determining the size of capacitors required for power factor improvement.

It was derived from the formulas:

$$pf = \cos \theta = kW/kVA$$

$$kVAr = kW \tan \theta$$

Scales A, B and D are used for problems involving kVA, pf and kW, and scales A, C and D are used for problems dealing with kVAr, pf and kW.

This nomograph can be used to solve a variety of power and power factor problems.

### EXAMPLE 1

What is the power factor when the system instruments indicate 100 kVA and 60 kW?

### Solution

Align 100 on scale A with 60 on scale D and read 0.6 power factor at the intersection on scale B.

### EXAMPLE 2

What kVAr value of capacitors must be installed on the system to raise the power factor from 60 to 80% when the plant load is 100 kW?

### Solution

Align 100 on scale D with 0.60 on scale C and read 132 kVAr on scale A. Now align 100 on scale D with 0.80 on scale C and read 74 kVAr on scale A. The difference between the two readings on scale A (132 – 74) is equal to 58 kVAr of capacitors required.

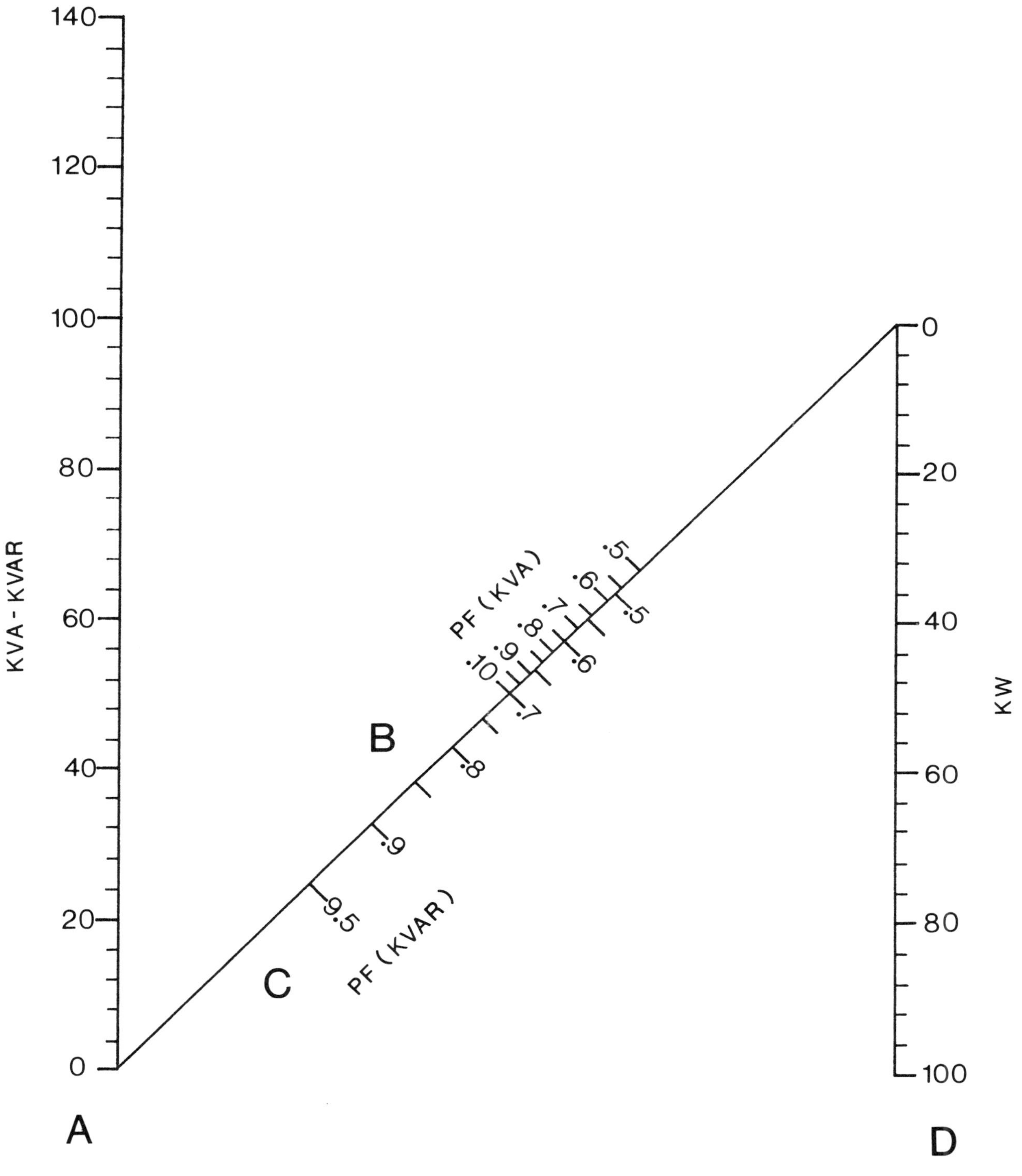
140
120
100
80
60
40
20
0
KVA - KVAR
A
B
C
PF (KVA)
PF (KVAR)
0
20
40
60
80
100
KW
D

## 3-14. Economics nomograph determines actual payback period

An important consideration for any equipment expenditure is the determination of the payback period. In other words, to determine how long it will take you to recover your investment.

With this nomograph you can quickly and easily determine the actual payback period. It is obtained from the following formula:

$$n = \log [1 + N(R - 1)/R]/\log R$$

where
$n$ = actual payback period, yr
$N$ = simple payback period, yr
= first cost, \$/savings per year, \$
$R$ = $(1 + e)/(1 + i)$
$e$ = escalation rate, %
$i$ = interest rate, %

### EXAMPLE

Determine the actual payback period in years when e = 10%, i = 12%, first cost = \$1000 and savings per year = \$200.

### Solution

Align 10% on scale e with 12% on scale i and read 0.984 on scale R. Align this point with 5 (\$1000/\$200) on scale N and read actual payback period on scale n as 5.3 yr.

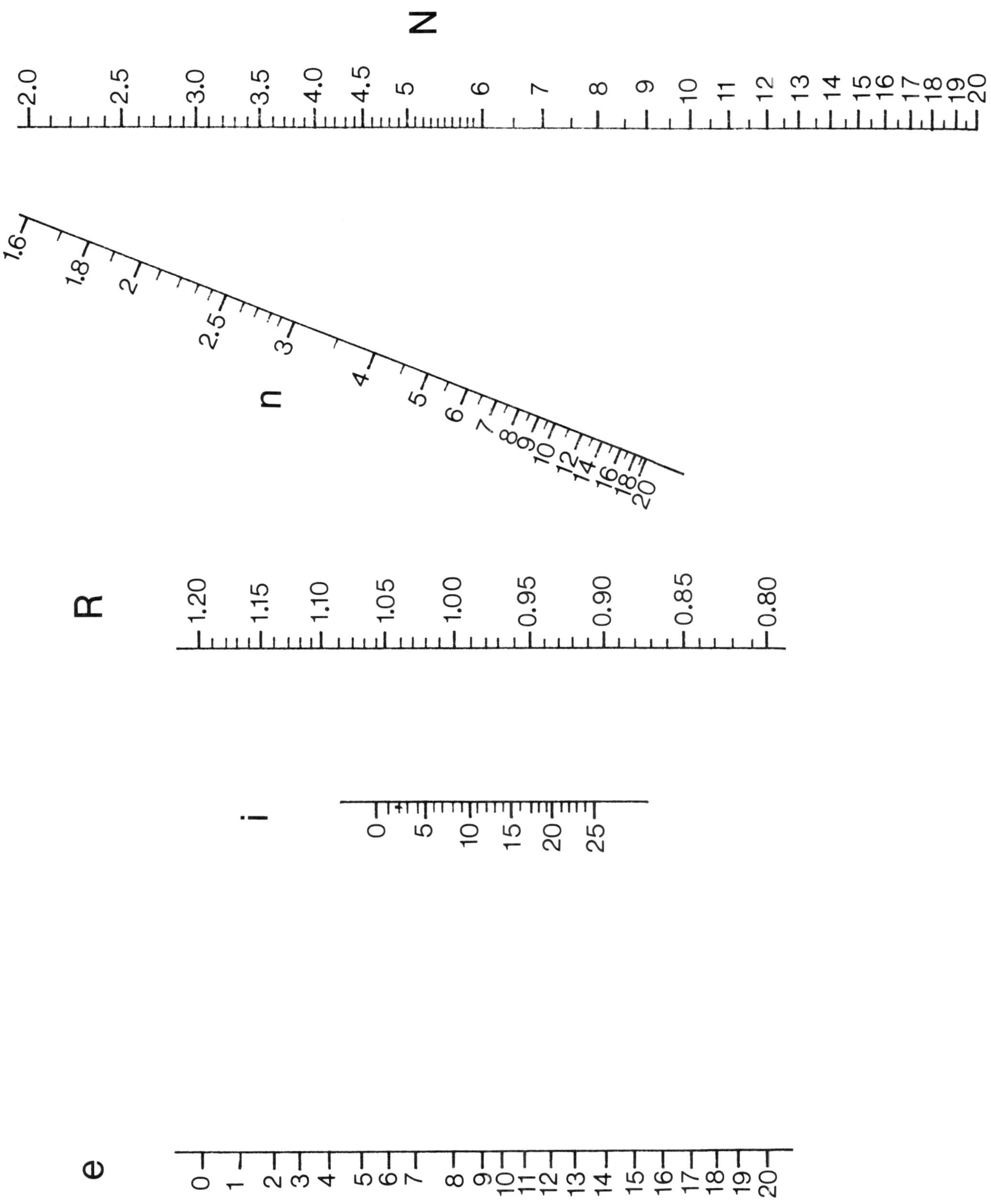
N
2.0 2.5 3.0 3.5 4.0 4.5 5 6 7 8 9 10 11 12 13 14 15 16 17 18 19 20
n
1.6 1.8 2 2.5 3 4 5 6 7 8 9 10 12 14 16 18 20
R
1.20 1.15 1.10 1.05 1.00 0.95 0.90 0.85 0.80
i
0 5 10 15 20 25
e
0 1 2 3 4 5 6 7 8 9 10 11 12 13 14 15 16 17 18 19 20

## 3–15. Bonus

With this handy nomograph you can quickly and easily determine the answers to various gasoline and operating cost–type problems. It is particularly helpful in determining rapidly and without any complicated and time-consuming mathematics cost comparisons for cars with different ratings in miles per gallon of gasoline. All you need is a straightedge and a couple of minutes of your time.

You may be pleasantly surprised to learn that you can trade your present gas-guzzler for a new high-mileage car for a dollar or so a day, considering your savings on gasoline.

Suppose your car is reasonably new, you drive the national average of 12,000 mi/yr and get 15 mpg. You would consume 800 gal of gasoline. The new car you have in mind gets 25 mpg and would use 480 gal. At $1.25/gal, the difference is $400/year. Would you pay about $140/month for the new car or would you keep your present car at a cost of about $30/month extra for gas?

### EXAMPLE

You drive an average of 15,000 mi/yr and your present car gets 15 mpg of gas. You have shopped around and found a car to your liking that is rated at 30 mpg. What is your saving, if you trade cars when gasoline costs $2.00/gal.

#### Solution

Align $2.00 on scale A with 15 mi on scale B and read 13.3¢ on scale C. Align 13.3¢ with 15,000 on scale D and read $1995 on scale E. This is your annual gasoline operating cost for your present car.

Repeat as above, align $2.00 on scale A with 30 mi on scale B and read 6.7¢ on scale C. Align 6.7¢ with 15,000 mi on scale D and read $990 on scale E.

Your cost savings in this case is more than $1000/yr. This yields a savings of $84/month, and that could be more than one-half your monthly payment on your new high-mileage car. In the future you may save even more, and you will be driving a new car.

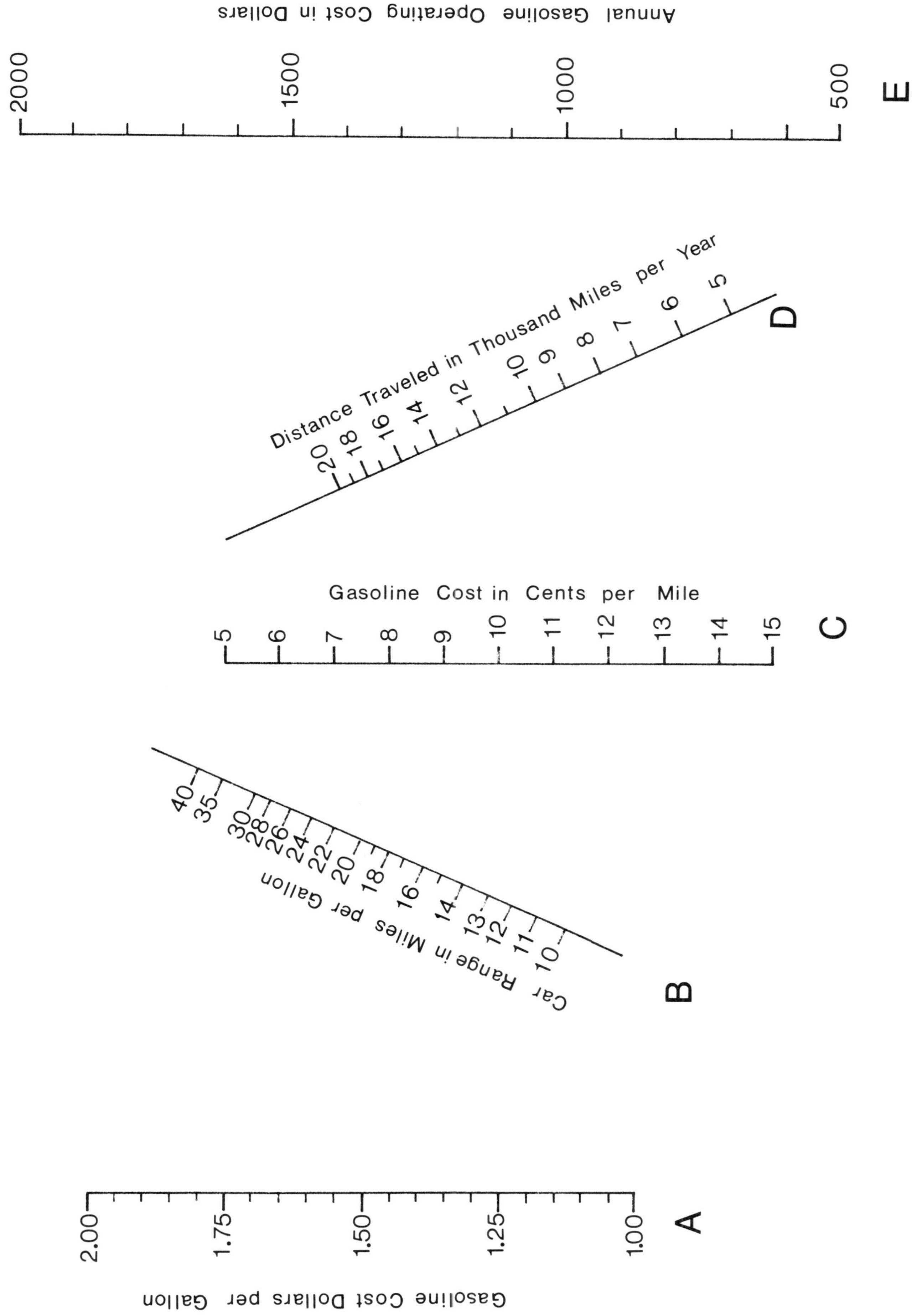

Annual Gasoline Operating Cost in Dollars
2000
1500
1000
500
E
Distance Traveled in Thousand Miles per Year
20
18
16
14
12
10
9
8
7
6
5
D
Gasoline Cost in Cents per Mile
5
6
7
8
9
10
11
12
13
14
15
C
Car Range in Miles per Gallon
40
35
30
28
26
24
22
20
18
16
14
13
12
11
10
B
Gasoline Cost Dollars per Gallon
2.00
1.75
1.50
1.25
1.00
A

# CREDITS

The author is grateful to the authors and publishers who kindly gave their permission to reproduce copyrighted material from the sources listed below.


1-1. Molnar, J. "Nomograph Converts acfm at Elevated Temperatures to scfm," *Heating/Piping/Air Conditioning* (November 1978), pp. 113–114.

1-2. Owens-Corning Fibreglas. "Properties of Moist Air," Psychrometric Chart, Engineering Data 10.E.6, Publication No. 4-in-3406, p. 6.

1-3. Shaw, W. J. D. "Nomograph Determines Moisture Addition, Removal Quantities," *Heating/Piping/Air Conditioning* (February 1974), pp. 53–54.

1-4. Molnar, J. "Nomograph Calculates Temperature of Mixed Air," *Heating/Piping/Air Conditioning* (March 1978), pp. 91–92.

1-5. Molnar, J. "Nomograph Calculates Required Ventilation Air Changes," *Heating/Piping/Air Conditioning* (November, 1975), pp. 63–64.

1-6. Molnar, J. "Determining Air Requirements of Exhaust Hoods, Slot Velocity," *Plant Engineering Magazine* (February 16, 1978), pp. 132–133, published by Technical Publishing, A Company of Dun & Bradstreet Corp.

1-7. Molnar, J. "Determining Air Requirements of Exhaust Hoods, Face Velocity," *Plant Engineering Magazine* (February 16, 1978), pp. 132–133, published by Technical Publishing, A Company of Dun & Bradstreet Corp.

1-8. Molnar, J. "Transport Air Velocity Calculator," *Pollution Engineering* (accepted for publication).

1-9. W. W. Sly Manufacturing Co. "Friction Loss Chart: Equivalent Resistance in Feet of Straight Pipe," Air Pollution Control Bulletin No. 204B, p. 32.

1-10. Molnar, J. "Nomograph Determines Parameters for HVAC Electric Duct Reheating," *Heating/Piping/Air Conditioning* (July 1976), pp. 87–88.

1-11. Molnar, J. "Calculating the Capacity of Electric Heating Coils," *Plant Engineering Magazine* (June 22, 1978), p. 127, published by Technical Publishing, A Company of Dun & Bradstreet Corp.

1-12. Industrial Engineering & Equipment Co. (INDEECO). "Calculating KW Requirements," Bulletin C20-1, p. 36.

2-1. Molnar, J. "Nomograph Gives Electric Water Heater Input," *Heating/Piping/Air Conditioning* (April 1979), pp. 103–104.

2-2. Molnar, J. "Nomograph Calculates Steam Required to Heat Water in Hot Water Converter," *Heating/Piping/Air Conditioning* (July 1978), pp. 79–80.

2-3. Molnar, J. "Nomograph Determines Parameters for Shell and Tube Heat Exchanger," *Heating/Piping/Air Conditioning* (May 1980), pp. 89–90.

2-4. Molnar, J. "Nomograph Determines Mean Velocity of Flowing Liquids in Pipes," *Heating/Piping/Air Conditioning* (March 1980), pp. 101–102.

2-5. Molnar, J., "Nomograph Determines Weight of Water-Filled Pipe," *Pollution Engineering* (accepted for publication).

2-6. Molnar, J. "Nomograph Determines Quantity of Water in Steel Pipes," *Pollution Engineering* (accepted for publication).

2-12. Molnar, J. "Quick Sizing of Building Drains," *Plant Engineering Magazine* (December 26, 1968), p. 50, published by Technical Publishing, A Company of Dun & Bradstreet Corp.

2-13. Molnar, J. "How to Determine Drain Pipe Size," *Plant Engineering Magazine* (August 8, 1968), p. 114, published by Technical Publishing, a Company of Dun & Bradstreet Corp.

2-14. Molnar, J. "Nomograph Determines Trench Excavation Quantities," *Heating/Piping/Air Conditioning* (February 1980), pp. 91–92.

3-1. Molnar, J. "Determining Power Loss in Conductors," *Electrical Construction & Maintenance* (September 1977), p. 124.

3-2. Molnar, J. "Determining Current-KVA Relationships," *Plant Engineering Magazine* (January 8, 1981), p. 73, published by Technical Publishing, A Company of Dun & Bradstreet Corp.

3-6. Molnar, J. "Determining Cost Savings for Equipment Shutdowns," *Pollution Engineering* (accepted for publication).

3-7. Molnar, J. "Finding Belt Velocity," *Plant Engineering Magazine* (December 13, 1979), p. 103, published by Technical Publishing, A Company of Dun & Bradstreet Corp.

3-8. Molnar, J. "Sizing Drive Pulleys," *Plant Engineering Magazine* (October 18, 1979), p. 173, published by Technical Publishing, A Company of Dun & Bradstreet Corp.

3-9. Molnar, J. "Duct Weight Calculator," *Pollution Engineering* 2(5):31 (November/December 1970).

3-10. Molnar, J. "Estimating Sheet-Metal Duct Costs," *Plant Engineering Magazine* (August 17, 1978), p. 145, published by Technical Publishing, A Company of Dun & Bradstreet Corp.

3-11. "Calculating Sound Levels for Fan Speed Changes," in *Plant Engineering Directory and Specifications Catalog* (Barrington, IL: Technical Publishing, A Compay of Dun & Bradstreet Corp., 1980), p. C–44.

3-12. "Calculating Sound Levels for Two or More Fans," in *Plant Engineering Directory and Specifications Catalog* (Barrington, IL: Technical Publishing, A Company of Dun & Bradstreet Corp., 1980), p. C–44.

3-14. "Economics Nomograph Determines Actual Payback Period," courtesy M. B. Herbert, PE.